BIG DATA WITH HADOOP MAPREDUCE

A Classroom Approach

BIG DATA WITH HADOOP MAPREDUCE

A Classroom Approach

Rathinaraja Jeyaraj
Ganeshkumar Pugalendhi
Anand Paul

Apple Academic Press Inc.
4164 Lakeshore Road
Burlington ON L7L 1A4
Canada

Apple Academic Press, Inc.
1265 Goldenrod Circle NE
Palm Bay, Florida 32905
USA

© 2021 by Apple Academic Press, Inc.

Exclusive worldwide distribution by CRC Press, a member of Taylor & Francis Group

No claim to original U.S. Government works

International Standard Book Number-13: 978-1-77188-834-9 (Hardcover)
International Standard Book Number-13: 978-0-42932-173-3 (eBook)

Library and Archives Canada Cataloguing in Publication

Title: Big data with Hadoop MapReduce : a classroom approach / Rathinaraja Jeyaraj, Ganeshkumar Pugalendhi, Anand Paul.
Names: Jeyaraj, Rathinaraja, author. | Pugalendhi, Ganeshkumar, author. | Paul, Anand, author.
Description: Includes bibliographical references and index.
Identifiers: Canadiana (print) 20200185195 | Canadiana (ebook) 20200185241 | ISBN 9781771888349 (hardcover) | ISBN 9780429321733 (electronic bk.)
Subjects: LCSH: Apache Hadoop. | LCSH: MapReduce (Computer file) | LCSH: Big data. | LCSH: File organization (Computer science)
Classification: LCC QA76.9.D5 .J49 2020 | DDC 005.74—dc23

CIP data on file with US Library of Congress

Apple Academic Press also publishes its books in a variety of electronic formats. Some content that appears in print may not be available in electronic format. For information about Apple Academic Press products, visit our website at **www.appleacademicpress.com** and the CRC Press website at **www.crcpress.com**

About the Authors

Rathinaraja Jeyaraj
Post-Doctoral Researcher, University of Macau, Macau

Rathinaraja Jeyaraj has obtained PhD from National Institute of Technology Karnataka, India. He recently worked as a visiting researcher at connected computing and media processing lab, Kyungpook National University, South Korea and supervised by Prof. Anand Paul. His research interests include big data processing tools, cloud computing, IoT, and machine learning. He completed his BTech and MTech at Anna University, Tamil Nadu, India. He has also earned an MBA in Information Systems and Management at Bharathiar University, Coimbatore, India.

Ganeshkumar Pugalendhi, PhD
Assistant Professor, Department of Information Technology,
Anna University Regional Campus, Coimbatore, India

Ganeshkumar Pugalendhi, PhD, is an Assistant Professor in the Department of Information Technology, Anna University Regional Campus, Coimbatore, India. He received his BTech from University of Madras, MS (by research), and PhD degrees from Anna University, India, and did his postdoctoral work at Kyungpook NationalUniversity, South Korea. He is the recipient of a Student Scientist Award from the TNSCST, India; best paper awards from IEEE, the IET, and the Korean Institute of Industrial and Systems Engineers; travel grants from Indian Government funding agencies like DST-SERB as a Young Scientist, DBT, and CSIR and a workshop grant from DBT. He has visited many countries (Singapore, South Korea, USA, Serbia, Japan, and France) for research interaction and collaboration. He is the resource person for delivering technical talks and seminars sponsored by Indian Government Organizations like UGC, AICTE, TEQIP, ICMR, DST and others. His research works are published in well reputed Scopus/SCIE/SCI journals and renowned top conferences. He has written two research-oriented textbooks: Data Classification Using Soft Computing and Soft Computing for Microarray Data Analysis. He is a Track Chair for Human Computer Interface Track in ACM SAC (Symposium on Applied

Computing) for 2016 in Italy, 2017 in Morocco, 2018 in France and 2019 in Cyprus. He is a Guest Editor for Taylor & Francis Journal and Inderscience Journal in 2017, Hindawii Journal in 2018, MDPI Journal of Sensor and Actuator Networks in 2019. His Citation and h-index are (260, 8), (218, 7) and (117, 6) in Google Scholar, Scopus and Publons respectively as on 2020. His research interests are in Data Analytics and Machine Learning.

Anand Paul, PhD
Associate Professor, School of Computer Science and Engineering, Kyungpook National University, South Korea

Anand Paul, PhD, is currently working in the School of Computer Science and Engineering at Kyungpook National University, South Korea, as Associate Professor. He earned his PhD in Electrical Engineering from the National Cheng Kung University, Taiwan, R.O.C. His research interests include big data analytics, IoT, and machine learning. He has done extensive work in big data and IoT-based smart cities. He was a delegate representing South Korea for the M2M focus group in 2010–2012 and has been an IEEE senior member since 2015. He is serving as associate editor for the journals *IEEE Access, IET Wireless Sensor Systems, ACM Applied Computing Reviews, Cyber Physical Systems* (Taylor & Francis), *Human Behaviour and Emerging Technology* (Wiley), and the *Journal of Platform Technology*. He has also guest edited various international journals. He is the track chair for smart human computer interaction with the Association for Computing Machinery Symposium on Applied Computing 2014–2019, and general chair for the 8th International Conference on Orange Technology (ICOT 2020). He is also an MPEG delegate representing South Korea.

A Message from Kaniyan

From Purananuru written in Tamil

யாதும் ஊரே யாவரும் கேளிர்
தீதும் நன்றும் பிறர்தர வாரா
நோதலும் தணிதலும் அவற்றோ ரன்ன
சாதலும் புதுவது அன்றே, வாழ்தல்
இனிதென மகிழ்ந்தன்றும் இலமே முனிவின்
இன்னா தென்றலும் இலமே, மின்னொடு
வானம் தண்துளி தலைஇ யானாது
கல் பொருது மிரங்கு மல்லல் பேரியாற்று
நீர்வழிப் படூஉம் புணைபோல் ஆருயிர்
முறை வழிப் படூஉம் என்பது திறவோர்
காட்சியில் தெளிந்தனம் ஆகலின், மாட்சியின்
பெரியோரை வியத்தலும் இலமே,
சிறியோரை இகழ்தல் அதனினும் இலமே. (புறம்: 192)

English Translation by Reverend G.U. Pope (in 1906)

To us all towns are one, all men our kin.
Life's good comes not from others' gift, nor ill
Man's pains and pains' relief are from within.
Death's no new thing; nor do our bosoms thrill
When Joyous life seems like a luscious draught.
When grieved, we patient suffer; for, we deem
This much – praised life of ours a fragile raft
Borne down the waters of some mountain stream
That o'er huge boulders roaring seeks the plain
Tho' storms with lightnings' flash from darken'd skies
Descend, the raft goes on as fates ordain.
Thus have we seen in visions of the wise !   (Puram: 192)

—Kaniyan Pungundran

Kaniyan Pungundran was an influential Tamil philosopher from the Sangam age (3000 years ago). His name Kaniyan implies that he was an astronomer as it is a Tamil word referring to mathematics. He was born and brought up in Mahibalanpatti, a village panchayat in the Thiruppatur taluk of Sivaganga district in the state of Tamil Nadu, India. He composed two poems called Purananuru and Natrinai during the Sangam period.

Contents

Abbreviations

ACID	Atomicity Consistency Isolation Durability
AsM	Applications Manager
AWS	Amazon Web Services
BR	Block Report
BSP	Bulk Synchronous Processing
CDH	Cloudera Distribution for Hadoop
CSP	Cloud Service Providers
DAG	Directed Acyclic Graph
DAS	Direct Attached Storage
DFS	Distributed File System
DM	Data Mining
DN	Data Node
DRF	Dominant Resource Fairness
DSS	Decision Support System
DWH	Data WareHouse
EMR	Elastic MapReduce
ETL	Extract Transform Load
FIFO	First In First Out
GFS	Google File System
GMR	Google MapReduce
GPU	Graphics Processing Unit
HA	High Availability
HB	HeartBeat
HDD	Hard Disk Drive
HDFS	Hadoop Distributed File System
HDT	Hadoop Development Tools
HDP	Hadoop Data Platform
HPC	High Performance Computing
HPDA	High Performance Data Analytics
HT	Hyper Threading
IDC	International Digital Corporation
IO	InputOutput
IS	Input Split

JHS	Job History Server
JN	Journal Node
JT	Job Tracker
JVM	Java Virtual Machine
LAN	Local Area Network
LHC	Large Hadron Collider
LSST	Large Synoptic Survey Telescope
MR	MapReduce
MRAppMaster	MR Application Master
MPI	Message Passing Interface
NAS	Network-Attached Storage
NCDC	National Climatic Data Centre
NFS	Network File System
NIC	Network Interface Card
NLP	Natural Language Processing
NM	Node Manager
NN	Name Node
NN HA	Name Node High Availability
NSF	National Science Foundation
NYSE	New York Stock Exchange
OLAP	OnLine Analytical Processing
P-P	Point to Point
QA	Question Answering
QJM	Quorum Journal Manager
RAID	Redundant Array of Inexpensive Disks
RDBMS	Relational Database Management System
RF	Replication Factor
RM	Resource Manager
RPC	Remote Procedure Call
RR	Record Reader
RW	Record Writer
SDSS	Sloan Digital Sky Survey
SAN	Storage Area Network
SNA	Social Network Analysis
SNN	Secondary Name Node
SPOF	Single Point of Failure
SPS	Stream Processing System
SSD	Solid State Disk
STONITH	Shoot the Other Node In the Head

TFLOPS	Tera Floating Operation Per Second
TT	Task Tracker
VM	Virtual Machine
WORM	Write Once and Read Many
WUI	Web User Interface
YARN	Yet Another Resource Negotiator
ZK	ZooKeeper
ZKFC	ZooKeeper Failover Controller

Preface

"We aim to make our readers visualize and learn big data and Hadoop Map Reduce from scratch."

There is a lot of content on Big Data and Hadoop MapReduce available on the Internet (online lectures, websites) and excellent books are available for intermediate level users to master Hadoop MapReduce. Are they helpful for beginners and non-computer science students to understand the basics of big data, Hadoop cluster setup, and easily write MapReduce jobs? This requires investing much time to read or watch lectures. Hadoop aspirants (once upon a time, including me) find difficulties in selecting the right sources to begin with. Moreover, the basic terminologies in big data, distributed computing, and inner working of Hadoop MapReduce and Hadoop Distributed File System are not presented in a simple way, which makes the audience reluctant in pursuing them.

This motivation sparked us to share our experience in the form of a book to bridge the gap between inexperienced aspirants and Hadoop. We have framed this book to provide an understanding of big data and MapReduce by visualizing the basic terminologies and concepts with more illustrations and worked-out examples. This book will significantly minimize the time spent on the Internet to explore big data, MapReduce inner working, and single node/multi-node installation on physical/virtual machines.

This book covers almost all the necessary information on Hadoop MapReduce for the online certification exam. We mainly target students, research scholars, university professors, and big data practitioners who wish to save their time while learning. Upon completing this book, it will be easy for users to start with other big data processing tools such as Spark, Storm, etc. as we provide a firm grip on the basics. Ultimately, our readers will be able to:

+   understand what big data is and the factors that influence them.
+   understand the inner working of MapReduce, which is essential for certification exams.
+   setup Hadoop clusters with 100s of physical/virtual machines.
+   create a virtual machine in AWS and setup Hadoop MapReduce.

+   write MapReduce with Eclipse in a simple way.
+   understand other big data processing tools and their applications.
+   understand various job positions in data science.

We believe that, regardless of domain and expertise level in Hadoop MapReduce, many will use our book as a basic manual. We provide some sample MapReduce jobs (https://github.com/rathinaraja/MapReduce) with the dataset to practice simultaneously while reading our text. Please note that it is not necessary to be an expert, but you must have some minimal knowledge of working in Ubuntu, Java, and Eclipse to setup cluster and write MapReduce jobs.

Please contact us by mail if you have any queries. We will be happy to help you to get through the book.

Dedication and Acknowledgment

This is an excellent opportunity for me to thank Prof. V.S. Ananthanarayana (my research supervisor, Deputy Director, National Institute of Technology Karnataka), Dr. Ganeshkumar Pugalendhi (my post-graduate supervisor in Anna University, Coimbatore), and Dr. Anand Paul (my research mentor in Kyungpook National University, South Korea) for being my constant motivation. I sincerely extend my gratitude to Prof. V.S. Ananthanarayana, for the freedom he provided to set my goals and pursue in my style without any restriction. It would not be an exaggeration to thank Dr. Ganeshkumar Pugalendhi and Dr. Anand Paul for their significant contribution in shaping and organizing the contents of this book more simply. It would not be possible to bring my experience as a book without their help and support. I am so much grateful to them. I want to thank Mr. Sai Natarajan (Director, Duratech Solutions, Coimbatore), Dr. Karthik Narayanan (Assistant Professor, Karunya University, Tamil Nadu), Mr. Benjamin Santhosh Raj (Data-Centre Engineer, Wipro, Chennai), Dr. Sathishkumar (Assistant Professor, SNS College of Technology, Coimbatore), Mr. Elayaraja Jeyaraj (Data-Centre Engineer, CGI, Bangalore), Mr. Rajkumar for spending their time to carry out the technical review, and Ms. Felicit Beneta, MA, MPhil, for language correction. I thank them for contributing helpful suggestions and improvements to my drafts. I also thank all who contributed regardless of the quantum of work directly/ indirectly. It is always impossible without the family support to invest massive time for preparing a book. I am debted to my parents, Mrs. Radha Ambigai Jeyaraj and Mr. Jeyaraj Rathinasamy, for my whole life. I am so much grateful to my brothers, Mr. Sivaraja Jeyaraj, and Mr. Elayaraja Jeyaraj, for supporting me financially without any expectation. Finally, I should mention my source of inspiration right from my graduate studies until research degree, my wife, Dr. Sujiya Rathinaraja, who consistently gave mental support all through the tough journey. Infinite thanks to her for keeping my life green and lovable.

—Rathinaraja Jeyaraj

Introduction

This book covers the basic terminologies and concepts of big data, distributed computing, and MapReduce inner working, etc. We have emphasized more on Hadoop v2 when compared to Hadoop v1 in order to meet today's trend.

Chapter 1 discusses the reasons that caused big data and why decision making from digital data is essential. We have compared and contrasted the importance of horizontal scalability over vertical scalability for big data processing. The history of Hadoop and its features are mentioned along with different big data processing framework.

Chapter 2 is built on Hadoop v1 to elaborate on the inner working of the MapReduce execution cycle, which is very important to implement scalable algorithms in MapReduce. We have given examples to understand the MapReduce execution sequence step-by-step. Finally, MapReduce weaknesses and solutions are mentioned at the end of the chapter.

Chapter 3 completely covers single node and multi-node implementation step-by-step with a basic wordcount MapReduce job. Some Hadoop administrative commands are given to practice with Hadoop tools.

Chapter 4 briefly introduces a set of big data processing tools in a Hadoop ecosystem for various purposes. Once you are done with Hadoop Distributed File System and MapReduce, you are ready to dirty your hands with other tools based on your project requirement. We have given many web links to download the various big datasets for practice.

Chapter 5 takes you into Hadoop v2 by introducing YARN. However, you will find it easy if you read Chapter 2 already. Therefore, we strongly recommend you to spend some time on Hadoop v1, which will help you to understand why Hadoop v2 is necessary. Moreover, the Hadoop cluster and MapReduce job configurations are discussed in detail.

Chapter 6 is a significant portion in our book that will explain Hadoop v2, single node/multi-node installation on physical/virtual machines, running MapReduce job in Eclipse itself (you need not setup a real Hadoop cluster to frequently test your algorithm), properties used to tune MapReduce cluster and job, art of writing MapReduce jobs, NN high availability, Hadoop Distributed File system federation, meta-data in NN, and finally creating Hadoop cluster in Amazon cloud. You will find this chapter more helpful if you wish to write many MapReduce jobs for different concepts.

Chapter 7 briefly describes data science and some big data problems in text analytics, audio analytics, video analytics, graph processing, etc. Finally, we have mentioned different job positions and their requirements in the big data industry.

The Appendixes includes various dataset links and examples to work out. We have also included a case study on NYSE dataset to have complete experience of MapReduce.

CHAPTER 1

Big Data

A journey of a thousand miles begins with a single step.
—Lao Tzu

INTRODUCTION

Big data has dramatically changed the way in businesses, management, and research sectors. It is considered to be an emerging fourth scientific paradigm called "data science." Let us have a quick review of the emergence of science over centuries.

Empirical Science – The proof of concept is based on experience and evidence verifiable rather than pure theory or logic.

Theoretical Science – The proof of concept is theoretically derived (Newton's law, Kepler's law, etc.) rather than conducting experiments for many complex problems, as creating evidence is difficult. It was also infeasible deriving thousands of pages.

Computational Science – Deriving equations over 1000s of pages for solving problems like weather prediction, protein structure evaluation, genome analysis, solving puzzle, games, human-computer interaction such as conversation was typically taking huge time. Application of specialized computer systems to solve such problems is called computational science. As part of this, a mathematical model is developed and programed to feed into the computer along with the input. This deals with calculation-intensive tasks (which are not humanly possible to calculate in a short time).

Data Science – Deals with data-intensive (massive data) computing. Data science aims to deal with big data analytics comprehensively to discover unknown, hidden pattern/trend/association/relationship or any other useful, understandable, and actionable information (insight/knowledge) that leads to decision making.

1.1 BIG DATA

New technologies, devices, and social applications exponentially increase the volume of digital data every year. The size of digital data created till 2003 was 4000 million GB, which would fill an entire football ground if piled up in disks. The same quantity was created in every two days in 2011, and every 10 minutes in 2013. This continues to proliferate. The data is meaningful and useful when processed. "**Big data** refers to a collection of datasets that are huge or flow large enough or with diverse types of data or any of these combinations that outpace our traditional storage (RDBMS), computing, and algorithm ability to store, process, analyze, and understand with a cost-effective way." How big "big data" is? In simple terms, any amount of data that is beyond storage capacity, computing, and algorithm ability of a machine is called big data. Example:

- 10 GB high definition video could be a big data for smartphones but not for high-end desktops.
- Rendering video from 100 GB 3D graphics data could be a big data for laptop/desktop machines but not for high-end servers.

A decade back, the size was the first, and at times, the only dimension that indicated big data. Therefore, we might tend to conclude as follows:

Big (huge) + data (volume + velocity + variety) → huge data volume + huge data velocity + huge data variety

However, the volume is one of the factors that chokes the system capability. Other factors can individually hold the neck of computers. Even though the last equation is true, volume, velocity, and variety need not be combined to say a dataset is big data. Anyone of the factors (volume or velocity or variety) is enough to say a field is facing big data problems if it chokes the system capability. From the definition, "big data" not only emerged just from storage capacity (volume) point of view, but also from "processing capability and algorithm ability" of a machine. Because hardware processing capability and algorithm ability determine how much amount of data a computer can process in a specified amount of time. Therefore, some definitions focus on what data is, while others focus on what data does.

Some interesting facts on big data

The International Digital Corporation (IDC) is a market research firm that monitors and measures the data created worldwide. It reports that

- every year, data created is almost doubled.
- over 16 ZB was created in 2016.
- over 163 ZB will be created by 2020.
- in today's digital data world, 90% were created in the last couple of years, in which 95% of data is in semi/unstructured form, and merely less than 5% belongs to structured form of data.

1.1.1   BIG DATA SOURCES

Anything capable of producing digital data contribute to data accumulation. However, the way data generated in the last 40 years has changed completely. For example,

before 1980 – devices were generating data.
1980–2000 – employees generated data as an end user.
since 2000 – people started contributing data via social applications, e-mails, etc.
after 2005 – every hardware, software, application generated log data.

It is hard to find any activity that does not generate data on the Internet. We are letting somebody else watch us and monitor our activities over the Internet. Figure 1.1 [1] illustrates what happened in every 60 seconds in the digital world in 2017 by Internet-based companies.

- YouTube users upload 400 hours of new video and watch 700,000 hours of videos.
- 3.8 million searches are done on Google.
- Over 243,000 images are uploaded, and 70,000 hours of video are watched on Facebook.
- 350,000 tweets are generated on Twitter.
- Over 65,000 images are uploaded on Instagram.
- More than 210,000 snaps are sent on Snapchat.
- 120 new users are joining LinkedIn.
- 156 Million E-mails are exchanged.
- 29 million messages, over 175,000 video messages, and 1 million images are processed in WhatsApp every day.
- Videos of 87,000 hours are watched on Netflix.
- Over 25,000 posts are shared on Tumblr.
- 500,000 applications are downloaded.

- Over 80 new domains are registered.
- Minimum of 1,000,000 swipes and 18,000 matches are done on Tinder.
- Over 5,500 check-ins are happening on Foursquare.
- More than 800,000 files are uploaded in Dropbox.
- Over 2,000,000 minutes of calls are made on Skype.

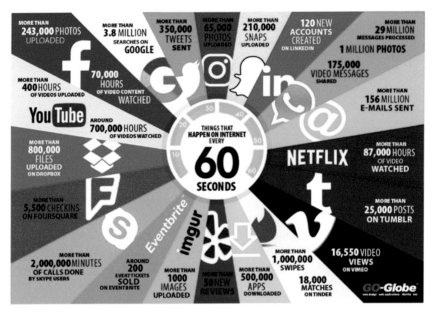

FIGURE 1.1   What is happening in every 60 seconds? (Reprinted from Ref. [1] with permission).

Let us discuss how big data impacts the major domains such as businesses, public administration, and scientific research.

Big data in business (E-commerce)

The volume of business data worldwide doubles every 1.2 years. E-com companies such as Walmart, Flipkart, Amazon, eBay, etc. generate millions of transactions per day. For instance, in 6000 Walmart stores worldwide,

- 7.5 million RFID entries were done in 2005.
- 30 billion RFID entries were accounted in 2012.
- exceeded 100 billion RFID entries in 2015.
- 300 million transactions are happening per nowadays.

Multimedia and graphics companies also face big data problems. For example, Avatar movie 3D video rendering required over 1 PB of local storage space, which is equivalent to 11,680 days long MP3. Credit card fraud detection system manages over 2 billion valid accounts around the world. IDC estimates that over 450 billion transactions will happen per day after 2020. New York Stock Exchange (NYSE) generates over 1 TB of trade data per day. Other industries like banking, finance, insurance, manufacturing, marketing, retail industries, pharmaceutical, etc. generate massive data.

Big data in public administration

In India, Aadhar is a unique identification number, and a large biometric database (over PB) recorded with every person's retina, thumb impression, photo, etc. The US Library of Congress collected 235 TB of data in 2011. People in different age groups need different public services. For example, kids and teenagers need more education; elders require health care services, etc. The government tries to provide high-level of public services with significant budgetary constraints. Therefore, they take big data as a potential budget resource to reduce national debts and budget deficits.

Big data in scientific research

E-science includes many scientific disciplines, where devices generate a massive amount of data. Most of the scientific fields have already become highly data-driven by the development of computer science. Example: astronomy, atmospheric science, earth science, meteorology, social computing, bioinformatics, biochemistry, medicine (medical imaging), genomics (genome sequencing), oceanography, geology, sociology, etc. produce a large volume of data with various types and different speed from different sources. Following are some of the scientific fields that are overloaded with data.

- National Science Foundation (NSF) initiated under ocean observatories and embedded fiber optic cable over 1000 km on the sea floor, connecting 1000s of biological sensors. It produces a huge amount of data to monitor environment behavior.
- Large Hadron Collider (LHC) is a particle accelerator device and produces 60 TB/day.
- National Climatic Data Centre (NCDC) generates over 100 TB/day.
- Large Synoptic Survey Telescope (LSST) is similar to a large digital camera that records 30 trillion bytes of images/day.
- Sloan Digital Sky Survey (SDSS) produces 200 GB images/day.

- A couple of decades back, decoding human genome took ten years to process.

Other sources

- IoT includes mobile devices, microphones, readers/sensors/scanners that generate data.
- CCTV for security and military surveillance, animal tracker, etc.
- Call logs analytics in the customer service center.
- Shortlisting CVs from millions of applications.
- Aircraft generate 10 TB black box data every 30 minutes, and about 640 TB is generated in one flight.
- Smart meters in power grids read the usage every 15 minutes and records 350 billion transactions in a year.

1.1.2   WHY SHOULD WE STORE AND PROCESS BIG DATA?

We started storing data, as storage device price became cheaper. We need historical data to learn, understand, and make decisions for adapting to changes and reacting swiftly in the future. The potential of big data is in its ability to solve problems and provide new opportunities. Same big data is processed multiple times for different purposes that give different insight from a different perspective. One of the active Facebook revenue sources is publishing ads on the page you like, share, comment. They allow E-commerce companies to access your data for money to publish ads. Some typical applications that have to handle big data are:

- Social Network Analysis (SNA): Social network data is rich in content and relationships that are quite valuable to many third-party business entities. They use such data for different purposes. For instance,
  - Understanding and targeting customers for marketing.
  - Detecting online communities, predicting market trends, etc.
- E-commerce: recommender system (people who like this product may also like another product in online shopping, friend suggestions on Facebook), sentiment analysis, marketing, etc.
- Banking and Finance: stock market analysis, risk/fraud management, etc.
- Transportation: logistic optimization, real-time traffic flow optimization, etc.

- Healthcare: medical record analysis, genome analysis, patient monitoring, etc.
- Telecommunication: threat detection, violence prediction, etc.
- Entertainment: animation, 3D video rendering, etc.
- Forecasting events like disease spread, natural disaster and take proactive measures.
- Optimizing system (hardware/software) performance.
- Improving performance in sports.
- Improving security and law enforcement.

Despite huge data trouble humans, there are so much potential and useful information hidden. Ultimately processing and extracting insight from big data should lead to

- increase productivity in business.
- improve operational efficiency, reduce risk, and make strategic decisions in management.
- ease scientific research.

1.1.3 BIG DATA CHARACTERISTICS (DIFFERENT V OF BIG DATA)

Any amount of data is big data when storage, computing, and algorithm ability fail to process and extract meaningful insight. Following are the indicators to mention big data.

1. Volume: *more data more accurate decisions*

Big data should not be discarded because the storage cost is cheap. We can derive more information from more data we store. Basic units of data measurement are shown in Table 1.1. Given a dataset, different users have different demands. Data that is not processed today may be worth enough when processed tomorrow. Relational models like Relational Database Management System (RDBMS) and Data Warehouse (DWH) cannot store and manage huge data in structured tables. To understand how information is measured and weighed, we have an interesting subject called "information science." There is a statement that "data growth exceeds Moore's law." What does it mean? In 1965, Moore stated that the number of transistors crammed into a processor chip doubles every 18 months resulting in doubling the processor performance. However, cramming more transistors into a chip turned out to be inefficient beyond a point due to excessive heat.

Later, processor technology was shifted to multi-core (2004) to increase processing speed in parallel.

TABLE 1.1   Data Size Units

In bytes	Unit	Binary	In bytes	Unit	Binary
1 Bit	0 or 1	-	1024 Kryat byte	1 Amos byte	2^{150} bytes
1 Byte	8 bits	2^{0} bytes	1024 Amos byte	1 Pectrol byte	2^{160} bytes
1024 Bytes	1 Kilo byte	2^{10} bytes	1024 Pectrol byte	1 Bolger byte	2^{170} bytes
1024 Kilo byte	1 Mega byte	2^{20} bytes	1024 Bolger byte	1 Sambo byte	2^{180} bytes
1024 Mega byte	1 Giga byte	2^{30} bytes	1024 Sambo byte	1 Quesa byte	2^{190} bytes
1024 Giga byte	1 Tera byte	2^{40} bytes	1024 Quesa byte	1 Kinsa byte	2^{200} bytes
1024 Tera byte	1 Peta byte	2^{50} bytes	1024 Kinsa byte	1 Ruther byte	2^{210} bytes
1024 Peta byte	1 Exa byte	2^{60} bytes	1024 Ruther byte	1 Dubni byte	2^{220} bytes
1024 Exa byte	1 Zetta byte	2^{70} bytes	1024 Dubni byte	1 Seaborg byte	2^{230} bytes
1024 Zetta byte	1 Yotta byte	2^{80} bytes	1024 Seaborg byte	1 Bohr byte	2^{240} bytes
1024 Yotta byte	1 Bronto byte	2^{90} bytes	1024 Bohr byte	1 Hassiu byte	2^{250} bytes
1024 Bronto byte	1 GeoP byte	2^{100} bytes	1024 Hassiu byte	1 Meitner byte	2^{260} bytes
1024 GeoP byte	1 Sagan byte	2^{110} bytes	1024 Meitner byte	1 Darmstad byte	2^{270} bytes
1024 Sagan byte	1 Pija byte	2^{120} bytes	1024 Darmstad byte	1 Roent byte	2^{280} bytes
1024 Pija byte	1 Alpha byte	2^{130} bytes	1024 Roent byte	1 Coper byte	2^{290} bytes
1024 Alpha byte	1 Kryat byte	2^{140} bytes			

But, why is big data regarded to Moore's law? Let us assume that 1 GB data is stored in Hard Disk Drive (HDD). A dual-core processor can finish processing the entire 1 GB data in parallel. Now, data size increased to 10 GB. Well, the number of cores in a processor also increased up to 8. Therefore, 10 GB data is processed very faster in parallel. Now, data size increased to 1 TB. Although it is possible to increase the number of cores to 16, 32, etc. in a processor chip, due to slow HDD InputOutput (IO) rate, most of the cores are idle without getting data from disks. Therefore, even the number of cores doubled in a processor, due to the slow HDD IO rate, it takes more time to process 1 TB.

Table 1.2 shows the lean improvement in the data transfer rate of HDDs over decades than memory and storage capacity growth rate. In today's multi-core processor trend, cores are idle most of the time due to poor IO rate. Therefore, a single system to process 1 TB of data may take a whole day or even more. Since data transfer rate of the HDD has not evolved much, there is no point in increasing the number of cores or number of processors on a motherboard to process huge amount of data. This is also called "CPU heavy, IO rate poor" in computer architecture.

TABLE 1.2 Data Transfer Rate in a Server Machine Over Decades

Year	Cores in a processor	Memory capacity	HDD capacity	Data transfer rate per second in HDD	Time to read entire drive (in minutes)
1990	1	32–128 MB	1 GB	4.5 MB	4.5 ~ 4
2010	2/4/8	16–64 GB	> 1 TB	100 MB	100 ~ 166
Performance					
	max read speed			**max write speed**	
HDD	122 MB/s			119 MB/s	
SSD	456 MB/s			241 MB/s	

Since 1990 till 2010, processor performance has increased due to multi-core technology (for parallel processing) and memory size has increased minimum 128 times. In this decade, the read and write performance improved using Solid State Disk (SSD) over 4x and 2x times compared to typical HDD as given below. Therefore, in the future, SSDs will be highly used in major business sectors. However, small scale businesses may not prefer SSDs as it is highly expensive per GB.

2. Velocity: *the faster more revenue*

Rate of incoming data to a machine for processing is faster and continuous, which demands the system to provide fresh, low latency results in real-time. Such continuous/streaming data is processed in the unit of windows (size in KBs). This streaming data must be processed in real time before persistently stored. RDBMS is not suitable for data velocity as it needs to index data before accessing. Real-time applications such as threat detection in telecommunication, fraud detection in banking, recommender system in social and e-commerce applications are the best use cases for data velocity.

3. Variety: *one tool to process different types of data*

Traditional data were just documents, logs, and transaction files. Nowadays, data are in different forms like audio, video, image, 3D, spatial, and temporal.

Structured data in RDBMS grows linearly in banking and other business sections. Unstructured and semi-structured data grow exponentially due to the growth of the Internet-based applications and IoT. Figure 1.2 [2] plots the growth of different types of data over the years. Every year unstructured data generation is doubled. Despite having enough computing facility, a data processing tool must have the capability to process different type of data.

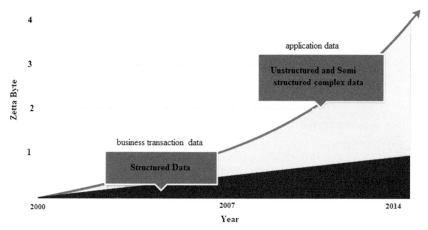

FIGURE 1.2   Growth of heterogeneous data. Adapted from Ref. [2] with permission.

Structured data: creating a table (schema) before you accumulate data is called structuring the data. It means that managing a collection of data in an organized manner using a frame (table). Changing schema after massive data accumulated is a time-consuming process in RDBMS.

Unstructured data: denotes data organization with dynamic schema or no schema. Example: Facebook users upload an image, video, audio, text (chat, status, comment). Such heterogeneous data cannot be stored in RDBMS as schema modification is costly for varying size data. Therefore, we go for NoSQL databases.

Semi-structured data: It is not as structured as relational databases. Example:

- Web pages have little structure with tags, but no restriction with the data inside the tags.
- Log data generated by machines/software/applications.
- Data in excel sheet.

Every person has one or more E-mail, Facebook, Twitter, and blog accounts. At least 10 MB of log data per day is generated while accessing them. This log data is processed by service providers to track user behavior for many purposes like a product recommendation, marketing campaign, etc. Imagine, if 1 million users generate 10 MB of log data in a day, it is roughly 10^6 x 10 MB (over 10 TB). Therefore, big data in any form is dangerous to handle. Similarly, if a web site generates 20 KB log data per day, 20+ million web sites generate (20 KB x 20x10^6) 400+ TB log data per day. The enterprise server's maximum capacity to read from disk is 100 MB/sec. Therefore, it takes a few days just to read weblog data. In summary,

> Structured data – banking, finance, business sections, etc.
> Semi-structured data – log data from hardware/software/applications, emails, web pages (XML/HTML), etc.
> Unstructured data – audio, video, and text documents, etc.

4. Value: *big data beats better algorithms*

It depends on the ability of algorithms to extract potential insight from any amount of data. Relevant information extracted from big data could be very less (see Figure 1.3) that questions the usefulness of the result. This requires some potential analytics to extract more insight from massive data to improve decision making.

5. Veracity: *uncertainty of accuracy and authenticity of data*

Data taken from public sources such as social networks may not be accurate most of the time, because the authenticity of users (anybody can post any data) is not reliable on the Internet.

6. Variability

Variability refers to the dynamic, evolving behavior of data generation sources.

7. Volatility

Volatility is determining how long the data is valid and how long it should be stored. This is very difficult to find from which point data is no longer relevant to the current analysis.

8. Complexity

Complexity means that unstable number of variables in a dataset and their high inter-connectedness. This troubles the processing tools and algorithms to extract precise knowledge.

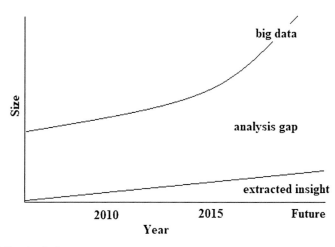

FIGURE 1.3   Analysis gap.

Big data processing is not about collecting data worldwide and processing them centrally. Every sector faces its own big data problems. A company may face any one of the Vs or a combination of any Vs. However, 'first three-V' typically exists in every firm. There are no universal benchmarks to define a range for volume, variety, and velocity. The defining limits depend upon the firm, and these limits evolve. It is also a fact that these characteristics may be dependent on each other. Table 1.3 summarizes the reasons for the different Vs emergence in short.

TABLE 1.3   Challenges for Computation and Algorithm Ability

Characteristics	
Volume	CPU heavy IO poor
Velocity	Demands more memory, CPU
Variety	The tool must support to process any type of data
Value	Need potential algorithm to extract more insight
Veracity	Uncertainty of data accuracy and authenticity
Variability	Dynamic and evolving behavior of data source
Volatility	Determines the data validity
Complexity	Unstable number of variables and its interconnectedness

Example 1: An insurance agency collects data about a person from various sources such as social media, bank transaction, web browsing activity, and decides whether to display him an insurance ad while booking travel ticket.

The agency considers its competitor's price and offers better value for attracting customers. Now, the insurance agency faces volume (historical data of customers, and their transactions), variety (data from social apps), and velocity (click streaming/current activity data), etc.

Example 2: General Motors (a company that manufactures and sells vehicles) has its data center facility to monitor running vehicle engine health. They predict the failure of their customer's vehicle engine and prepare for replacement before the vehicle owner arrives. This will undoubtedly increase the customer experience. Moreover, they sell such information to the insurance agencies for deciding insurance claim according to the speed they drive. Therefore, an accident by the rash drive cannot claim insurance.

1.1.4   DATA HANDLING FRAMEWORK

Our desktop system fumbles even to load a single word file of size 100 MB by using the local file system. Table 1.4 distinguishes the size of data and tools used to handle them.

TABLE 1.4   Data Handling Framework

Class	Size	Tools	How it fits
Small	<10 GB	Excel, R, MATLAB	Hardly fits in one computer's memory
Medium	10 GB – 1 TB	DWH	Hardly fits in one computer's storage disk
Big	>1 TB	Hadoop, Distributed DB	Stored across many machines

1.2   BIG DATA SYSTEMS

The potential of big data analytics is in its ability to solve business problems and provide new business opportunities by predicting trends. In E-commerce, applications like ads targeting, collaborative filtering (recommendation system), sentiment analysis, marketing campaign are some of the use cases that require to process big data to stay upright in business and increase revenue. There are two classes of big data systems:

- Operational big data system (Real-time response from big data)
- Analytical or decision support big data system (Batch processing)

1.2.1   OPERATIONAL BIG DATA SYSTEM

It is essential to understand the difference between transactional databases and operational databases to proceed further.

Transaction: A transaction is a set of coordinated operations performed in a sequence. Example: money transfer involves deducting money from one account and adding it to another account dealing with more than one table in a database. These operations must be performed in a sequence to ensure consistency.

Transactional database: A database that supports transactions for day-to-day operations is called a transactional database. Transactional databases are highly structured and heavily used in banking, finance, and other business applications. Example: RDBMS.

Operational database: A database that does not support transactions, but still performs day-to-day operations (such as insert, update, delete) is called operational database. It is highly used in web-based applications such as Facebook, WhatsApp, etc. NoSQL databases are called operational databases. Example: MongoDB, Big Table (Google), Cassandra (Facebook), HBase, etc. NoSQL databases do not support transactions because synchronization (to ensure consistency) limits the scalability of distributed systems. MongoDB tries to support both transactional and operational functionalities.

Both transactional and operational databases try to respond in real-time to the user. However, transactional databases are write-consistent, and operational databases are read-consistent. Some NoSQL databases provide a few analytical functions to derive insights from data with minimal coding effort, without the need for data scientists and additional infrastructure. NoSQL big data systems are designed to take the advantages of cloud computing to adopt massive scalable computing and storage inexpensively and efficiently. This makes operational big data workloads much easier to manage.

1.2.2   ANALYTICAL OR DECISION SUPPORT BIG DATA SYSTEM

Big data analysis refers to a sequence of steps (capture, store, manage, process, perform analytics, visualize/interpret, and understand) carried out to discover unknown hidden pattern/trend/relationship/associations and other useful information from big data for decision making using statistical, probability, data mining and machine learning algorithms in a cost-effective way. This is also called a decision support system and highly relies on batch processing that takes minutes to hours to respond. These include systems like

- DWH that stores structured historical data provided predefined queries like Online Analytical Processing (OLAP).
- Distributed storage such as Hadoop Distributed File System (HDFS), S3, Azure blob, Swift, etc. to store semi/unstructured data with no predefined queries. So, users have to write algorithms (ad-hoc algorithm) to process huge data.

These two classes of big data systems are complementary and frequently deployed together. Table 1.5 differentiates these two classes more precisely.

TABLE 1.5   Operational vs Analytical Big Data Systems

Characteristics	Operational	Analytical
Latency	1–100 ms	1 – 100 minutes
Concurrency	1000–100,000 users	1 -10 users
Access pattern	Write, read, update	Initial load, read, no update
Queries	Selective	Batch fashion
End-user	Customer	Data scientist
Technology	NoSQL	DWH, Hadoop, Spark

1.3   PLATFORM FOR BIG DATA PROCESSING

What do we do when there is too much data/computation to process? Scale the system. The ability of a system to adapt more resources (CPU, memory, storage) to tackle the increasing amount of data/computation is as called as scalability. Traditionally, scaling computer systems and algorithms are being the most significant challenges to tackle the increasing volume of data. There are two types in scaling the system, as shown in Figure 1.4: scale up and scale out. It is very crucial to choose the right platform for the right problem. The decision depends on how fast you want the result from a given dataset.

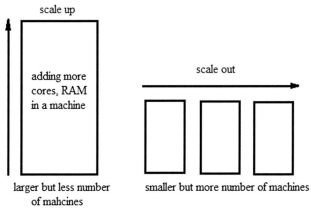

FIGURE 1.4   Scale out vs scale up.

1.3.1   HORIZONTAL SCALABILITY (SCALE OUT)

Resizing the cluster by adding multiple computers that work together as a single logical machine is called horizontal scalability or scale-out architecture. As we keep adding machines, we need to distribute computation/data across many servers to process in parallel. It can be scaled out on the fly by adding servers while the cluster is up and running. This is highly suitable for off-line (batch) processing. Scale-out is more challenging as it requires software that runs in a distributed environment to handle fault-tolerance. However, hardware costs and software licensing are cheap. Example: Hadoop, Spark, etc.

1.3.2   VERTICAL SCALABILITY (SCALE UP)

Resizing a computer by adding more processors and memory is called vertical scalability or scale-up architecture. However, scaling up reached its limit imposed by hardware. It is suitable for real-time processing. Adding resources to a server cannot be done on the fly as the system needs a reboot to detect the newly added hardware. Therefore, system downtime is unavoidable. Example: Graphics Processing Unit (GPU). Scale up is more expensive than scaling out. Scale up leads to CPU heavy IO poor problem. Therefore, there is no use of increasing resources of a machine beyond a point for big data processing (but fruitful for compute-intensive tasks). That is the reason, supercomputers and GPUs are not mostly preferred for big data processing.

Moreover, the failure of anything in such systems is costly to treat (needs downtime) and not affordable for most of the small-scale companies.

1.3.3   HORIZONTAL VS VERTICAL SCALING FOR BIG DATA PROCESSING

From Figure 1.5, when load increases in a cart, will you make your horse grow bigger or use multiple horses [15]? Of course, more donkeys. Similarly, when data size increases beyond the storage capacity of a computer, we go for tying up many computers than adding more storage drives in a server. Then, big data is divided into several pieces and stored in multiple computers. Therefore, big data can be processed in parallel.

FIGURE 1.5   Vertical or horizontal scale?

One significant advantage of using scale-out architecture is, failure of one computer does not halt the entire cluster, and replacing a computer is affordable. At run time, we can increase/decrease the number of nodes linearly and dynamically without cluster downtime. Table 1.6 accounts for the differences between these two types of system scaling. Following calculation shows the performance difference of using the single machine and a cluster of 10 machines to read 1 TB of data with 4 IO channels with bandwidth (transfer rate) of 100 MB/s as shown in Figure 1.6.

TABLE 1.6   Horizontal vs Vertical Scalability

	Advantages	Disadvantages
Horizontal scaling	It increases the performance in small unit linearly and can scale out the system as much as needed on the fly.	It is complex to build software to provide data distribution, fault-tolerance, and handle parallel processing complexities.
	We can use commonly available hardware, and software is opensource.	A limited number of software is available.
Vertical scaling	Most of the software can easily exploit the advantages of vertical scaling	Hardware and software costs are high. CPU heavy IO rate poor problem exists. So, scale up after a limit is not possible. It requires downtime (reboot) to scale up.

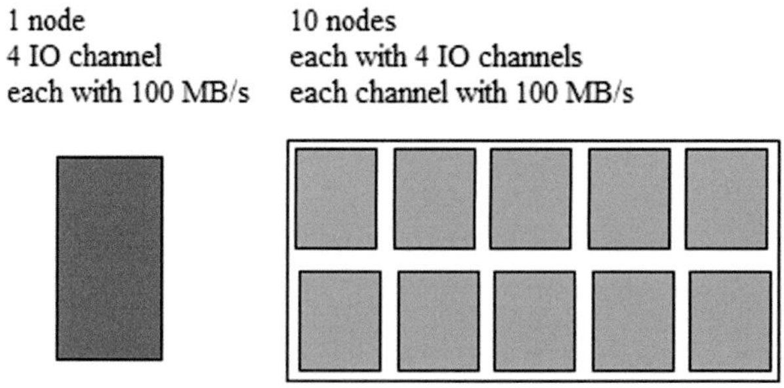

1 node
4 IO channel
each with 100 MB/s

10 nodes
each with 4 IO channels
each channel with 100 MB/s

FIGURE 1.6   Reading 1 TB of data with single vs multiple nodes.

Using a single machine =

$$\frac{1\,TB}{4 \times 100 \text{ MB}/\sec} = \frac{2^{10}}{400 \times 2^{20}/\sec} = \frac{2^{20}}{400}\sec = 2621.44 \sec = 43.69 \text{ minutes}$$

Using a cluster of 10 machines, each storing part of big data, say 102.4 GB =

$$\frac{1\,TB}{10 + 4 \times 100 \text{ MB}/\sec} = \frac{2^{40}}{4000 \times 2^{20}/\sec} = \frac{2^{20}}{4000}\sec = 262.144 \sec = 4.369 \text{ minutes}$$

Now, which one would you prefer? One computer or a cluster of computers? Cluster computing to handle present data deluge. To get the performance of $10,000 worth server for data processing, we can use four computers worth of $500 each as shown in Figure 1.7.

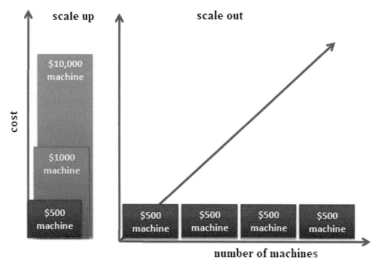

FIGURE 1.7   Cost of scale out and scale up.

1.4   EXISTING DATA PROCESSING FRAMEWORK: MOVE DATA TO THE NODE WHICH RUN PROGRAMS

Recent trends in data-center architecture have moved towards standardization and consolidation of hardware (storage, and network) to cut down the expenses. First, let us understand the drawbacks of the existing data-center architecture for data processing with an example (vote counting) to proceed further. Figure 1.8 illustrates that the votes from respective states are moved to Delhi for counting. Then, the result is sent back to those states. This involves a lot of network bandwidth to transfer votes and takes more time to compute centrally with one or a few machines.

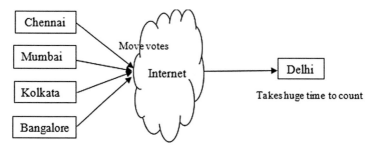

FIGURE 1.8   Centralized vote counting.

Similarly, traditional data management and processing system in data-center use infrastructure, in which compute cluster (racks of servers, also called nodes), and storage cluster (racks of storage drives) are physically separate as shown in Figure 1.9. Compute, and storage clusters are connected with high-speed network/copper/fiber optic cables. All data are stored in a centralized storage cluster like Network-Attached Storage (NAS). One or more servers in the compute cluster run analytical tools like DWH for data analysis and operational databases for real-time response. So, data from the storage cluster has to be brought into any one of the nodes running DWH or any other analytical tools to carry out an analysis.

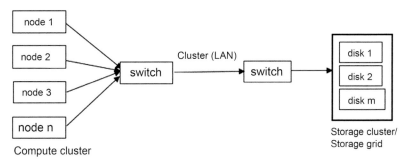

FIGURE 1.9   Existing data analytics architecture in data-center.

Since all servers in a compute cluster run the same software, this data processing architecture is highly available, fault-tolerant, load balancing, and scalable. The storage cluster uses a Redundant Array of Inexpensive Disks (RAID) to provide fault-tolerance for data and provides a file abstraction. As shown in Figure 1.10, merely less than 5% of overall collected data is moved from the storage cluster to compute cluster after performing ETL

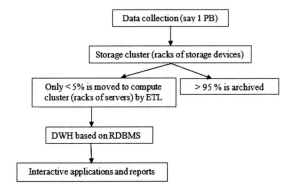

FIGURE 1.10   Existing ETL based data analytics flow.

(Extract-Transform-Load) for analysis. Rest, 95% of data, is archived in the storage grid. Archived data may have more meaningful insights but are not processed yet. Relational DWH tools such as Informatica, Teradata, Exadata, Syncsort, PowerCenter, Cognos, etc. are used to build and manage structured DWH. Many companies are using this ETL based processing architecture.

However, there are some drawbacks in the traditional distributed system that limit the performance of existing data processing architecture.

- Scalability was limited and not linear. This means that as you increase the number of servers in the compute cluster, the overall performance decreases due to synchronization and coordination.
- Huge network bandwidth was required in between compute and storage cluster. This caused a lot of processing power spent on transporting data.
- Expensive servers were needed, and scaling up/down was not a smooth process.
- Chances of hardware failure was high as 1000s of nodes are in the cluster. Therefore, loss of data and computation were unavoidable and difficult to handle such partial failure.

1.5 ADVANCED DATA PROCESSING FRAMEWORK: MOVE THE PROGRAM TO THE NODE WHICH HAS DATA

Continuing with vote counting example, to avoid huge data transfer over the network and reduce data processing time in the destination, votes are counted in the respective states and sent the summary to Delhi for a final decision as shown in Figure 1.11.

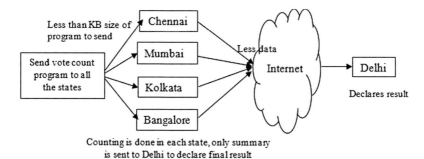

FIGURE 1.11 Distributed processing.

Moving data from the storage cluster to compute cluster involves huge local network bandwidth. Therefore, rather than managing storage and compute cluster separately, deploy only compute cluster, in which each machine is having its storage as shown in Figure 1.12. So, the collected data is divided and stored in different servers. Each server in the cluster manages a portion of huge data. To perform an analysis, send the program to the node which has the required data and get the results back.

This consumes very less network bandwidth as we distribute small size of the program (few KBs) rather than sending data. Therefore, no data is archived, and analytics can be done on the entire data in parallel every time. If the cluster storage capacity is exhausted out and some more data need to be loaded, include additional nodes (scale out) instead of increasing the storage capacity of the machine. Moreover, data from one node can be moved to another node running DWH using ETL for analysis, as shown in Figure 1.13.

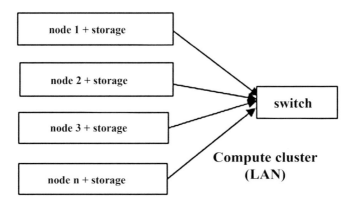

FIGURE 1.12   Shared nothing architecture.

Similarly, when your data increased to huge size (say 10 TB) that you cannot manage in your laptop or desktop, approach a data-center (cloud) and upload your data. In order to process them, you have to download it to your local machine. This cost a lot of public bandwidth and takes huge time to process in a small computer. So, create a program locally and send it to the data-center for processing data remotely. Cloud data-center also processes your data and sends the results back. This way, managing huge personal data is simplified using cloud computing.

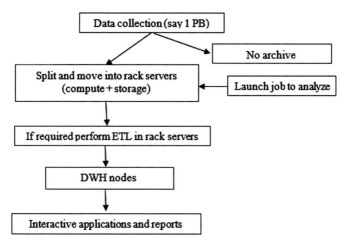

FIGURE 1.13 Data-parallel architecture.

1.5.1 ADVANTAGES OF ADVANCED DATA PROCESSING FRAMEWORK

Along with network bandwidth consumption minimization and processing time reduction, there are other benefits as well using advanced data processing framework.

- Distributed storage provides fault-tolerance using replication, which eliminates the requirement of RAID.
- Distributed processing helps to achieve data locality as the program is sent to the server which has stored the desired portion of data.
- Scalability (almost linear): as you increase the cluster size, performance increases.
- Low-cost hardware is enough.
- No licensing fees.

1.5.2 CHALLENGES IN THE HORIZONTAL FRAMEWORK

Horizontal scalability is robust only when there is no blocking and synchronization. In a large cluster,

- computers fail every day.
- data is corrupted or lost.
- computation is disrupted.

- the number of nodes in a cluster may not be constant.
- nodes can be heterogeneous.

Solving these problems using Message Passing Interface (MPI) for scale-out architecture is possible. However, it is the programer's responsibility to distribute data/program, failure handling, managing the cluster, networking, etc. It is challenging to build reliability into each application/job. It drastically tires programers and reduces productivity. Therefore, common infrastructure and a standardized set of tools are required to handle these complexities. It should be distributed, scalable, fault-tolerant, efficient, and easy to use.

1.5.3   REQUIREMENTS OF A DISTRIBUTED SYSTEM FOR ADVANCED DATA PROCESSING FRAMEWORK

A typical distributed system demands the following requirements in order to function more efficiently and effectively and for smooth application development.

1. Fault-tolerance – ability to recover data, task, and node from failure.
2. Consistency – node failure during the execution of a job should not affect the job outcome.
3. Scalability – adding nodes should result in a linear increase in performance.
4. Partial failure – failure of a node should result in a graceful degradation of application performance, but not the failure of the entire system.
5. Node recovery – if a failed node is repaired, it should be able to rejoin the cluster.

1.6   HADOOP

Let us summarize the traditional big data processing framework and converge into Hadoop.

1. Consider you have 1 TB HDD in your machine.
   Case 1: Can you store 300 GB of data? Yes. Can you store 700 GB of data? Yes. However, processing (read from disk, perform analysis) 1 TB data using a machine may take more than a day. Therefore, it demands more processing capacity to complete processing faster.

However, beyond a point, you cannot scale up resources due to "CPU heavy and IO rate poor" problem. Therefore, even high-end servers like supercomputer and GPU are not cost-effective to process huge data.

Case 2: Can you store 2 TB of data? No. It is beyond the storage capacity (1 TB) of a machine.

Solution: In the above two cases, the only way is to use a cluster of servers and employ a divide and conquer algorithm. Divide huge data into chunks, store in multiple servers in the cluster, process them in parallel with the same program, and combine the result.

2. Previously, the compute cluster and storage cluster were separate. Applications were built with MPI to process huge data. Computation was processor bound; in a sense, we had to send data to the node where a particular program is running and collect back the result. It was highly network-intensive and gave its worst for increasing data volume.

   Solution: Store data in the local storage of servers in compute cluster (so need not use storage cluster). Move computation (program) to a node which has the required data and process the data there itself locally without moving over the network. This is called data locality/ data gravity and massively helps to achieve horizontal scalability because the network bandwidth is consumed very less as program size is tiny (few KBs).

3. Cheap nodes fail, especially if you have many. So, programming using MPI for distributed systems is challenging to ensure fault-tolerance. Moreover, distributing code/data, networking, handling data/program/node failure, file handling, etc. should be taken care of by the programer for each application separately.

   Solution: Need a common framework that handles all the programer responsibilities of a distributed system and lets programers write only application-level algorithms.

To achieve the solutions proposed above, we have a software-based framework called Hadoop, which is an opensource project comprising a broad set of tools to store and process big data in a distributed environment across a cluster of commodity servers using a simple programming model. It is designed to scale out from single node to thousands of nodes, each offering its processing power and storage capacity. The challenge is not about storing massive data, but it is about the processing speed and developing distributed algorithms. Hadoop is highly distributed, horizontally scalable, fault-tolerant, high throughput, flexible, and cost-effective software solution for big data processing. As Hadoop is an opensource software and runs on commodity hardware, any small-scale company can use and modify the codes to their requirement. However, a big data solution is just more than Hadoop. Our primary focus is to derive insight from huge data using data mining and machine learning algorithms for decision making with the help of the Hadoop framework.

1.6.1   HISTORY OF HADOOP

For many years, users stored data in a database and processed via SQL queries for analysis. The DWH was used to store structured historical data and process them using OLAP for knowledge extraction. Relational databases could not store unstructured data and lacked scalability also. Before Hadoop, data in DWH was sampled/pre-processed before performing analysis due to lack of computing facility. Google, a web search engine monster, faced problems in processing huge data. Google wanted to download the whole Internet and index to support search queries with MySQL database. Google indexed 1 million pages in 1998, one billion in 2008, and over a trillion pages every day after 2010. So, Google implemented Google File System (GFS), Google Map Reduce (GMR), and Big Table in C++ for large-scale index data processing. Later, Doug Cutting and Mike Cafarella implemented HDFS and MapReduce in java based on GFS and GMR. Scalability in a distributed system and algorithm to support huge data processing in the web search engine was the key driving factor for Hadoop development. The brief history is,

> 1997 – NASA researchers faced big data problems.
> 1998 – Google was founded.
> 1999 – Apache software foundation was established.
> 2000 – Lucene: Text search library (by Doug Cutting).
> 2002 – Nutch project was started. It is an opensource web search engine in Yahoo and required hardware of $500,000 with monthly running cost of $300,000. They could achieve scalability just up to 20–600 nodes (Dreadnaught project).
> 2003 – GFS white paper was released.
> 2004 – Nutch distributed file system was designed based on GFS.
> 2004 – GMR white paper was released.
> 2005 – Nutch MapReduce was designed based on GMR.

> 2006 – Nutch distributed file system + Nutch MapReduce together renamed as Hadoop (as a subproject of Nutch) and separated from the Nutch project.
> 2008 – Hadoop was handed over to Apache software foundation, renamed as Apache Hadoop and opensourced.
> 2009 – Apache Spark
> 2011 – Apache Storm
> 2012 – Yet Another Resource Manager (YARN)

Doug Cutting named his software from his family interests. There is no relation between the project name and its functionality. The names are not an acronym too. Example:

> Lucene – his wife middle name.
> Nutch – a word used by his son to denote "meal."
> Hadoop – Yellow stuffed elephant named by his son.

1.6.1.1 *HADOOP SOFTWARE RELEASE*

The software goes through many versions from the first release until it becomes stable.

Alpha release – It means that an application is working, but some functionalities are likely to be missing. Several known and unknown bugs are likely to surface.

Beta release – It is a version that has been tested internally and being tested by the wider community. The software in the beta phase will generally have some bugs in completed software, speed/performance issues, and may cause crash or data loss.

Stable release – It is a version that is entirely tested and ready for use. Its functionality, specification, API are considered 'final' for that version. Apart from security patches and bug fixes, the software will not usually change for many years.

You must always use stable release for Hadoop installation. The standard release version is named as "x.y.z." Example: Hadoop 1.2.1, Hadoop 2.7.0, etc.

- x stands for a major release (may break compatibilities with the previous version)

- y stands for minor release (backward compatible)
- z stands for point release (backward compatible)

Hadoop 1.x.x is on API upgrades while Hadoop 2.x.x goes through architectural upgrades.

1.6.1.2   OS THAT SUPPORTS HADOOP

Almost all the OS such as Linux (Ubuntu, CentOS, Open Suse, Red Hat), Solaris, Windows, Mac, etc. support Hadoop. There may be some changes in the installation procedure for different OS. Moreover, the binary installation file also is different for each OS.

1.6.1.3   HADOOP DISTRIBUTIONS

All commercial Hadoop distributions are the modifications of original Apache Hadoop with a nice GUI. A lot of online service providers offer Hadoop service and support over the Internet for money. These distributors have their version of Hadoop with unique features. They also provide a Virtual Machine (VM) with pre-installed tools such as HDFS, MapReduce, Pig, Hive, HBase, Oozie, etc. Such VM is called quick start VM, which means that all the tools have already been installed and is ready for use. We need to download and install in our machine. Notable distributors are:

1. Apache Hadoop
   - Manual installation of Hadoop and its ecosystem such as pig, hive, etc.
   - No commercial support, only open forum discussions.
   - BigTop project manages and deals with packaging and testing Hadoop ecosystem.
2. Cloudera – Cloudera Distribution for Hadoop (CDH)
   - Apache Hadoop stack-based implementation for production.
   - Cloudera manager tool helps easy deployment, management, and monitoring.
3. Hortornworks – Hortornworks Data Platform (HDP)
   - Only distribution without any modification in Apache Hadoop.
4. MapR

- MapR includes Network File System (NFS) instead of HDFS in its distribution.
- Supports native UNIX file system.

5. Greenplum

1.6.2 HADOOP FEATURES

In simple words, Hadoop transforms a cluster of commodity servers into a service that stores and processes PBs of data reliably in a cost-effective way. Hadoop is capable of handling volume and various problems in big data. It is also offered as a cloud service (HDInsight from Microsoft Azure, Elastic MapReduce (EMR) from Amazon, etc.). Many companies have their flavor of Hadoop by modifying Apache Hadoop opensource code. The most specific features of the Hadoop framework are given below.

- **Highly distributed**

Hadoop works on a distributed file system and distributed computing. So, data and program can be moved around in a compute cluster.

- **Horizontally scalable**

Primary challenges for big data processing are the scalability of cluster and algorithms. RDBMS is scalable (limited) but not linearly scalable. Hadoop framework supports horizontal scalability for compute cluster and facilitates to write scalable algorithms using MapReduce. Hadoop supports AP in CAP theorem [16] for a distributed system. Losing consistency in CAP theorem, shared nothing architecture, moving computation to data, no blocking and synchronization among tasks allow Hadoop more horizontally scalable. Theoretically, there is no maximum limit in scaling out. Moreover, the application program need not be rewritten according to scaling.

- **Fault-tolerance**

Cheap nodes fail, especially if you have many in the cluster. Meantime between failures for a server is three years. Meantime between failures for 1000 servers is about one day. So, the failure of servers in a cluster is more likely. Hence, data and computation loss are apparent. Fault-tolerance is the ability of a system to recover from failure automatically and remain functional. Hadoop self-corrects (autonomic computing) when data/process gets lost. Hadoop provides fault-tolerance at the software-level. That is, Hadoop itself has been designed to detect and handle failures. However, why not hardware-level fault-tolerance? Anything hardware assisted is costly and prone to failure. So, managing resources like CPU, storage, and the network became software-defined.

- **High throughput**

Throughput is the amount of data processed per unit of time. As you increase the number of nodes in the Hadoop cluster, throughput increases linearly (which is not true in RDBMS) as shown in Figure 1.14.

- **Flexible software**

When a Hadoop cluster is up and running, you can dynamically add/remove nodes on the fly without disturbing/shutting down the cluster.

- **Commodity hardware**

Hadoop is designed to run on commodity servers. This means that you are not tied to expensive, proprietary hardware from a single vendor. You can use standardized, commonly available hardware from any vendor to build your Hadoop cluster. The commodity does not mean our laptop/desktop machines, which are cheap and has a higher failure rate. Nonetheless, Hadoop can be deployed in our personal computers/laptop too, but running them 24/7 is not possible.

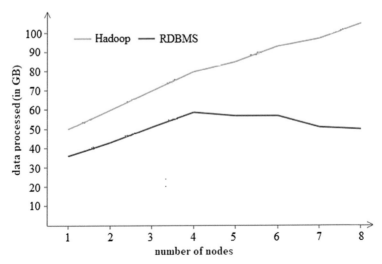

FIGURE 1.14   Hadoop vs RDBMS.

- **Rack-aware/topology-aware**

Hadoop is fed the network topology (arrangement of nodes in a cluster) via a configuration file, using which Hadoop can decide where to place data to minimize network flow and improve fault-tolerance.

- **Batch processing centric**

Batch processing means collecting data for some time and processing them together. It needs input data loaded before launching jobs. A batch-processing job has to read the dataset from the beginning till the end as there is no random read support. The same job may give varying latency in different runs. Moreover, typical batch processing jobs take more time (minutes to hours), so these jobs are not interactive, and users need not wait for the result after job submission.

- **Shared nothing cluster**

One server in a cluster cannot access the resources (memory, storage) of another server. Therefore, each server in Hadoop cluster autonomously functions.

Finally, Hadoop application developers need not worry about networking, file IO, distributing program and data, parallelization and load balancing, task/data/node failure (fault-tolerance). Hadoop takes care of these complexities and lets programers concentrate on writing algorithms to process the data. Hadoop completely hides system-level complexities from programers.

1.6.3   IS HADOOP A SOLUTION FOR ALL TYPES OF APPLICATIONS?

Although Hadoop is highly appreciated for solving major engineering problems that deal with big data, it is not suitable for some instances and applications. Hadoop is not

- for transaction processing (RDBMS is better for transactions). Because, the index is given only for data blocks, and not for the content available in it.
- for low latency processing (for every job, it has to process entire big data).
- good for a large number of small files (because of the size of meta-data increases).
- fair for compute-intensive applications that involve blocking, synchro-nization, and communication among processes.

- suitable for interactive, iterative, and stream processing.
- giving multiple writers, so file writing is continued at the end of file (appending), and no arbitrary (random) read/write possible.
- good when work cannot be parallelized.

Hadoop is optimized to

- process massive amounts of data in parallel.
- process structured/unstructured/semi-structured data.
- run on an inexpensive commodity server.
- scale out cluster very flexible to increase storage and computation power linearly.

1.6.4   COMPANIES USING HADOOP

Almost every business and non-profit sector face big data problem which hinder their productivity and development. Therefore, they are forced to embrace big data processing tools to improve their solution finding. Following are some of the major companies who already use the Hadoop framework to handle big data problems.

- Google uses Hadoop to build web index for Google search, and cluster articles for Google news.
- Yahoo uses Hadoop to detect spam in yahoo mail and construct a web map for search engine.
- Facebook makes use of Hadoop more extensively for applications such as ad optimization, spam detection, etc.
- Hadoop is also highly used by many universities and research labs: Astronomical image analysis (Washington), Bioinformatics (Maryland), Analyzing Wikipedia conflicts (PARC), Natural language processing (CMU), Particle physics (Nebraska), Ocean climate simulation (Washington).
- Head to https://wiki.apache.org/hadoop/PoweredBy for more information.

1.7   BASIC TERMINOLOGIES

Before discussing different big data processing frameworks and tools, let us understand some basic confusing terminologies that are interchangeably used.

Technique is a way to solve a problem and it is also typically termed as algorithm. A set of algorithms may be possible to solve the same problem.

Tools are programming languages used to implement techniques and perform any specific activities.

Technology is the scientific and engineering knowledge employed for building applications into usage and easing the task.

Framework/system consists of a stack/set of functions/components/tools working together and interact with each other to accomplish a common goal. Each function is called a layer/black box providing an abstraction, which means that each layer knows only its input, not the functionality of other layers, and may communicate with the layers above and below. Moreover, the failure of one layer does not affect the other layer in the system. Such a concept is implemented in many known software stacks. Example: OSI, DBMS, etc.

Now, let us see how these terminologies are used in big data environment with some examples.

Big data techniques – Any algorithm such as data mining for extracting unknown hidden pattern and machine learning for deriving actionable knowledge from big data is typically termed as big data technique.

Big data tools – help to implement big data techniques and perform ETL activities. Example: HDFS, MapReduce, Pig, Hive, Spark, Storm, etc.

Big data technologies – Data is processed differently, as its nature and behavior vary depending on applications. Example: batch processing, stream processing, in-memory computing, NoSQL, etc.

Big data framework – Big data framework provides an infrastructure for a set of big data tools to work together. Example: Hadoop, Spark, Stratosphere, Storm, etc.

Let us understand the differences among data analysis, data analytics, and data processing in terms of big data.

Data analytics is a systematic way of applying logic in data mining and machine learning algorithms using statistics/probability/mathematics to extract knowledge that helps in decision making. For example, data of a list of student's heights in cm is given {163, 154, 123, 145}. Finding average height (146.25 cm – knowledge extraction) is simple data analytics. With average height, one can decide the average shirt size of these students (decision making). Applying logic on big data that resides in a cluster of servers is called big data analytics. Traditional data mining and machine learning algorithms work incrementally for data stored in one server, but not suitable for data stored in a cluster of servers. Therefore, we need to modify these algorithms using big data-parallel programming tools such as MapReduce, Spark, Pig, Hive, etc. to work with data residing in a cluster of servers and discover unknown hidden pattern/trend/relationship/associations and other useful information from big data for decision making in a cost-effective way.

Data analysis is a careful study of an entity and presents the details of it. It involves a sequence of steps (capture/collect, store, manage, process, perform analytics, visualize/interpret, and understand) for the decision-making process. It is a multi-faceted process that involves different approaches and diverse techniques depending on the type of data and the purpose of the analysis. Data analytics is one of the sub-components in data analysis. Big data analysis deals with data analysis on a large scale.

Big data processing is a process of handling a large volume of information. Handling means working with the tools used in the sequence of steps in big data analysis.

1.8   DATA PROCESSING TECHNOLOGY

Data is processed in different fashion depending upon the data source and data type. There are various tools used for a different purpose in data processing technology. Notable tools are mentioned in Table 1.7.

TABLE 1.7   Flavors of Data Processing

Data processing technology	Tools	Big data handled
Batch processing	MapReduce, Spark	Volume, Variety
Stream processing (real-time processing)	Strom, S4, SEEP, D-stream, Samza	Velocity
In-memory processing (for interactive, iterative and stream processing)	Spark-SQL, D-stream	Volume, Variety, Velocity

1.9   BIG DATA FRAMEWORK

A big data framework consists of a set/stack of tools to deploy and work together for building scalable applications to process data differently for different purposes. Typical big data frameworks are Hadoop, Spark,

Stratosphere, etc. Different big data tools are used for different purposes such as storage, resource management, execution engine, etc. as shown in Figure 1.15. The order of big data tools deployment is essential. For instance, the HDFS is a primary tool to deal with file system operations. In order to perform data analytics using MapReduce, firstly, we have to install HDFS to store and retrieve big data. Similarly, MapReduce should have been already installed to work with Pig and Mahout. Therefore, each level in big data framework indicates the order of tool installation one after another. Let us discuss the purpose of each level in a big data processing framework.

Storage

Traditional file system software (FAT, NTFS, EXT3) is not designed for large-scale data processing because efficiency has a higher priority than other features. The massive size of data tends to store across multiple machines in a distributed way. Example: HDFS, GFS, Amazon S3, etc.

Database

RDMS was not designed to be distributed (but comes with huge cost due to synchronization and coordination). NoSQL databases use BASE properties by relaxing one or more ACID properties to achieve distributed, scalable storage. Different data model (key/value, column-family, document, graph) is used in NoSQL databases because of its flexible nature, unlike relational databases. Example: Dynamo, Voldemort, Riak, Neo4J, BigTable, HBase, Cassandra, MongoDB.

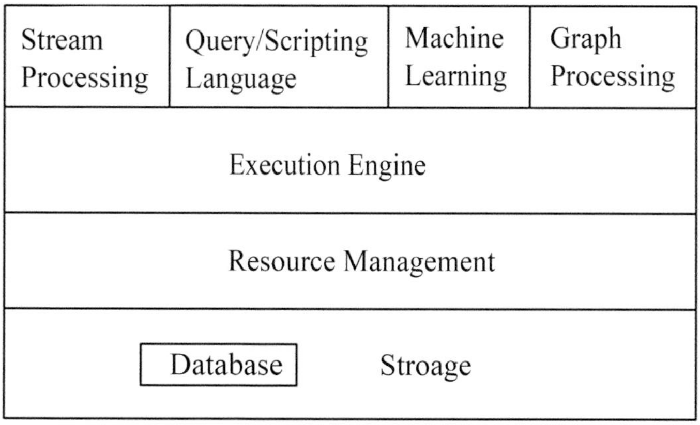

Stream Processing	Query/Scripting Language	Machine Learning	Graph Processing
Execution Engine			
Resource Management			
Database Stroage			

FIGURE 1.15   Big data analytics stack.

Resource management

In general, every big data framework is deployed in a dedicated cluster because different frameworks require different computing resources. This results in resource under-utilization. Large organizations need the ability to share data and resources of a cluster among multiple frameworks with clear isolation. Some of the resource management tools are Mesos, YARN, Quincy, etc.

Execution engine

It is a parallel programming tool to develop distributed and scalable algorithms on a cluster of unreliable, commodity servers. Typical execution engine tools are MapReduce, Spark, Stratosphere, Dryad, Hyracks, etc. It must be highly fault-tolerant and simple to use.

Query/scripting language

Low-level programming of execution engine such as MapReduce is not easy for non-java developers. So, we need a high-level language to improve the query capabilities of execution engines. Example: Pig, Hive, Shark, Meteor, DryadLinq, Scope, etc. 100 lines of Java code for MapReduce can be written in 10 lines using Pig scripts. A non-java programer can directly use Pig/Hive without knowing MapReduce functionalities. However, these high-level Pig/Hive scripts are translated into MapReduce API at run time.

Stream Processing System (SPS)

Processing incoming data before persistently stored and providing users a fresh, low latency result are called streaming processing. Since data is processed in the memory itself, there is no need for indexing mechanism as shown in Figure 1.16. Example: Storm, S4, SEEP, D-Stream, Naiad, Samza, Kafka, etc.

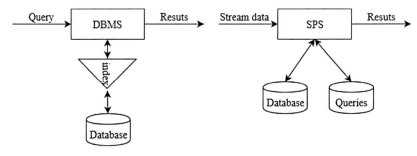

FIGURE 1.16   DBMS vs SPS.

Graph processing

Many real-world problems can be expressed as graphs and solved using graph algorithms. However, it involves significant computational dependencies, and multiple iterations to converge. A data-parallel framework such as MapReduce is not ideal for these problems because it involves large disk IO, which makes processing too slow. There are already some specialized graph processing libraries, which can be readily used for graph-based problems. Example: Pregel, Giraph, GraphX, GraphLab, PowerGraph, GraphChi, etc.

Machine learning

Implementing and consuming machine learning techniques at scale are difficult for developers and end users. So, there exist libraries that provide scalable data mining and machine learning algorithms. Example: Mahout, MLBase, SystemML, Ricardo, Presto, etc.

1.10   DIFFERENT BIG DATA PROCESSING FRAMEWORK

Some of the big data processing frameworks and the tools used in it are shown in Figure 1.17, Figure 1.18, and Figure 1.19.

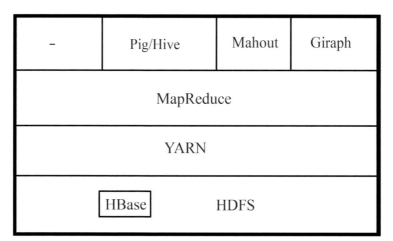

FIGURE 1.17   Hadoop.

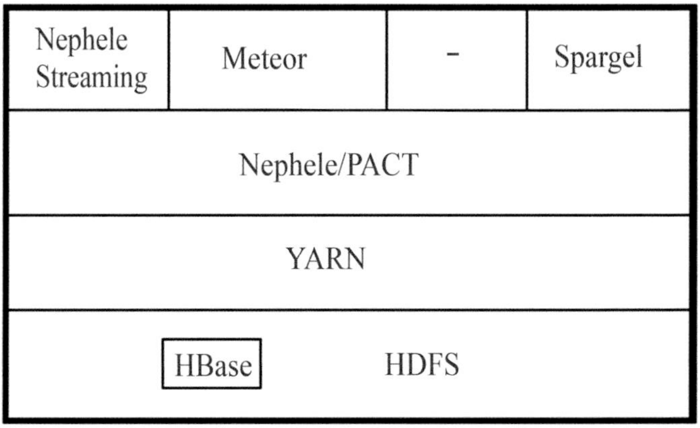

FIGURE 1.18   Stratosphere.

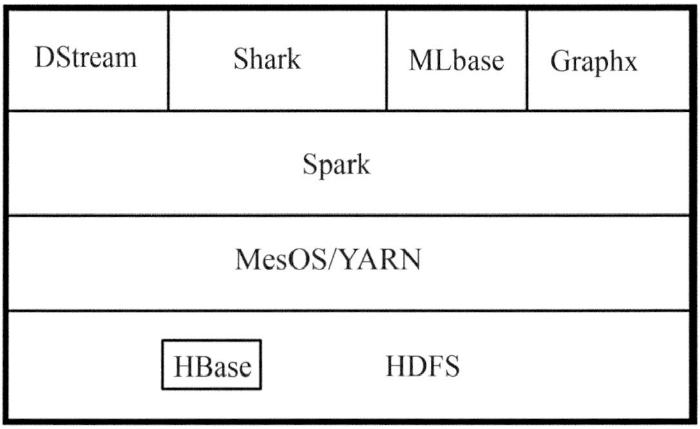

FIGURE 1.19   Spark.

How to choose a framework and tool for a big data problem? The decision depends on

- data size – If data fits into single system memory, then the cluster is not required. GPU or multicore CPU can be used in this case. If data does not fit in single system memory, then go for Hadoop, Spark, etc.
- throughput – If batch processing is needed, prefer any one of the scale-out architectures such as Hadoop, Spark, etc.
- data speed – If a real-time response is required, then choose a scale-up architecture such as Storm, Spark, etc.

For machine learning algorithms:

- Training an algorithm to find a prediction model requires to scale out framework such as Hadoop, Spark, etc. that runs off-line. But why off-line? Because machine learning algorithms need to process many input data, so it takes some time to learn. If machine learning algorithms are trained online, the learning process of the algorithms is disturbed by other processes.
- Once a prediction model is found, it is employed to predict future data in real-time. This requires to scale-up a framework to respond faster.

1.11   RDBMS VS HADOOP

Hadoop completely differs from RDBMS and supports huge scalability. As already explained, we need big data tools for the file system, resource management, execution engine in order to process data. All these components are built-in with RDBMS and work only with storage devices attached within a server. In big data frameworks, all these components are separate tools and installed independently in different servers. Other technical differences between Hadoop and RDBMS are.

1. Seek time in disks improved slower than the data transfer rate of disks. Seek time is the time taken to move read/write head to the particular place in the disk, whereas transfer rate denotes the bandwidth of the disk or amount of data that can be transferred per second. If seek time is higher, bandwidth is wasted. In traditional databases, B-tree is limited by seek time. This is avoided in HDFS as it indexes data blocks, not the content inside the data block.
2. RDBMS stores only structured data, whereas Hadoop (HDFS) tends to store any type of data (image, video, text) as it is just a file copy.
3. Data in HDFS is not normalized. For example, a hostname in a server log file repeats multiple time. That is the reason, log files are typical dataset to be processed by Hadoop.
4. RDBMS does not support the speed of data coming in.

Table 1.8 gives you additional differences between these two data processing tools. However, relational databases started incorporating analytical functions, and Hadoop started focusing on providing low latency results (Hive, HBase) for interactive access. Therefore, the line between these tools is becoming a blur.

Doubling the data size with doubling the cluster size is faster in Hadoop. This is not true in RDBMS. Because scaling is too costly, and achieving synchronization across the cluster is difficult. Similarly, you can think about why we cannot store huge structured database in multiple machines rather than finding a distributed model like Hadoop. The reason is that synchronization is a major problem for scaling out in relational databases. Table 1.9 summarizes the differences between distributed database and Hadoop.

TABLE 1.8   RDBMS vs Hadoop

Characteristics	RDBMS	Hadoop
Data size support	Up to TB	Over PB
Data Access	Interactive and batch	Only batch
	Transactional and operational	Operational and analytical
	Supports random access	No random access
Query	Supports point queries as records are indexed.	The only block is indexed not the content inside the block.
	Read, write, and update a small quantity of data many times.	Write once, read many times, and no update is allowed.
Data	Structured	Any data
Structure	Static schema	No schema or dynamic schema
	Schema on write	Schema on reading
Redundancy	Often normalized	Need not be normalized
Constraints	Strongly enforced	No constraints, or minimal constraints on write.
Indexing	Many indexing schemes available.	Minimal index support
Integrity	High (ACID)	Low (using checksum)
Scalability	Non-linear	Linearly scalable
Data locality	Move data to a server running particular database software.	Move code/program to the node having required data
Query response time	Immediate	High latency
Data distribution	Comes with the high cost	Easy and less cost
Framework	Need enterprise-level server, costly	Commodity server is enough, economical

TABLE 1.9 Distributed Database vs Hadoop

Characteristics	Distributed database	Hadoop
Computing model	Distributed query transaction	Distributed processing
	Ensures consistency and concurrency control.	No consistency and concurrency control required.
Cost	Need expensive servers.	Commodity servers are good enough.
Data model	Structured data	Any data
	Read, write, and update	Write once and read many times
Fault-tolerance	Failures are rare, RAID is used to recover.	Failures are common, software-defined fault-tolerance is used.
Key features	Efficiency, query optimization	Horizontal scalability, fault-tolerance

1.12 TRADITIONAL DWH VS HADOOP

This is very important to understand how OLAP and traditional data mining algorithms work and why they do not perform well to process huge amount of data. Firstly, you cannot apply data mining and OLAP on transactional databases because they serve millions of users in real-time. Moreover, about 50% of data in transactional databases may not be relevant for decision making. Therefore, we go for DWH to maintain structured historical data taken from transactional databases over some time.

The DWH is a subject oriented, integrated, time variant (using a timestamp) and non-volatile huge relational database. We perform ETL to bring data from transactional databases deployed geographically in different locations. Once the initial load is done, we find aggregation using predefined queries like OLAP or ad-hoc data mining algorithms to extract a pattern from DWH for decision making. There are some unique differences between OLAP and data mining, as shown in Table 1.10. However, what if my data is unstructured like text, image, etc. that cannot be loaded into DWH? Moreover, OLAP has no predefined queries to extract patterns from unstructured data.

A **data mart** is a simple form of DWH that is focused on a single subject. A data mart is often built and controlled by a single department within an organization. Given their single-subject focus, data mart usually draws data from only a few sources. Significant differences between DWH and data mart are shown in Table 1.11.

While performing ETL to load data into DWH, we remove some data from it. Data is generated and transferred at a considerable cost, so we must completely store without losing tiny data. Later, we can perform any analytics. Table 1.12 differentiates traditional DWH with big data infrastructure.

TABLE 1.10   Data Mining vs OLAP

Data mining algorithms	OLAP in DWH
Data driven	User driven, and verification driven
Need to pre-process (ETL) the dataset before apply algorithms.	Need to pre-process (ETL) the data to load into DWH itself.
Finds the unknown hidden pattern and then perform analytics.	Finds aggregation and makes a decision. Set a hypothesis, if it is met, take action, else no hypothesis.
Data mining algorithm can be applied in DWH, structured/unstructured data.	OLAP can be applied only in relational DWH.
It is user-defined ad-hoc algorithms.	It is a predefined query for relational DWH.
Gives you a pattern, which may or may not convey information.	Gives you only aggregated value.

TABLE 1.11   DWH vs Data Mart

Characteristics	DWH	Data mart
Scope	Corporate	Single department
Location	Centralized	Distributed
Subject	Multiple	Single subject
Data sources	Many	Few
Size	100 GB -TB+	< 100 GB

TABLE 1.12   Traditional DWH vs Big Data Infrastructure

Characteristics	Traditional DWH	Big data infrastructure
Data source	Transactional databases	Data from any source
Data structure	Predefined schema	Unstructured in nature
Consistency	High	Very low
Data location	Centralized	Physically distributed
Data analysis	Batch-oriented	Batch, stream, real-time, interactive
Cost	Requires high-end hardware and licensed software	Inexpensive commodity hardware and opensource software
CAP theorem	Consistency is a top priority.	Availability is a top priority.
Amount of data to process	Only the required data is taken for processing using ETL.	Entire data is processed.
Data model	Built on top of the relational data model.	Most of the big data databases are based on columnar databases.
Data Load	ETL	ELT
Example	DWH, data mart, data vault	HDFS, S3, HBase
Queries	OLAP, Ad-hoc algorithms	Ad-hoc algorithms

1.13 HPC VS GRID VS CLUSTER FOR BIG DATA PROCESSING

A processor having a large number of cores (GPU) or a cluster of nodes performing a huge unit of FLOPS is called High-Performance Computing (HPC). Typical HPC runs on an infrastructure, where storage and compute cluster are separate, as shown in Figure 1.9. The HPC uses a shared file system (Storage Area Network (SAN)/NAS) for the storage cluster. It uses MPI for communication among different processes executing in different nodes in the compute cluster. To achieve data processing, it has to bring data blocks from the storage cluster to compute cluster. So, cluster bandwidth becomes a bottleneck, therefore it uses shared memory architecture.

Managing cluster, detecting computation/data loss, checkpointing and recovery, network failure, etc. are the programer's responsibility. The major drawback is, moving data from storage cluster to compute cluster, which is expensive as cluster bandwidth is competitive among nodes. However, the HPC is predominant for compute-intensive tasks that need over TFLOPS. To minimize data movement across the local network, **Hadoop** co-locates data with compute nodes. Each compute node in the cluster has storage attached and stores part of huge data. Program code is sent to the node which has the required data, called data-locality or data gravity. It uses shared nothing architecture to achieve scalability, as shown in Figure 1.12. This architecture can be used for HPC too, but latency will be high.

Grid computing is a cluster of loosely coupled heterogeneous computers that are geographically located and work together to accomplish a task. Grid is a formation of a cluster of clusters and leverages shared nothing architecture. Grid computing is formed either as a physical machine cluster or a virtual machine cluster. One of the applications of grid computing is volunteer computing. To process huge amount of data, a small chunk of it is sent to a user who is part of the Internet and has the willingness to share his computing power. This is called volunteer computing. There are many volunteer computing projects: SETI, BOINC, etc. Hadoop seems similar in concept with volunteer computing as it performs distributed parallel processing. However, there are some unique differences:

- Volunteer computing runs on the untrusted Internet while Hadoop works on LAN (dedicated cluster).
- Volunteer computing breaks huge data into small chunks (0.3 MB in SETI) and sends to more than one user on the Internet to avoid failure, and fake result finally combines the output from volunteers. However, sending huge data to volunteers is not possible as users will donate CPU but not his bandwidth.
- Volunteer computing is suitable for compute-intensive jobs while Hadoop solves data-intensive problems.
- Volunteer's connection speed on the Internet around the globe is not the same. Hadoop needs high aggregated bandwidth that is available in LAN.

1.14 METRICS

It is essential to understand some standard metrics frequently used in data analysis. We have given a one-line description of each metric in Table 1.13.

TABLE 1.13   Basic Metrics and Its Description

Metric	Description
Response time	It is the difference between the submission time of a job/task and the time at which it starts executing.
Execution time	It is the amount of time a job/task spends actively with CPU. This is also a running time or CPU time.
Latency	The time taken by a job/task since its submission until its completion is called latency. This includes the execution time along with intermediate delay. Latency is a terminology used in networking. This is also called as processing time as job resides in memory or turnaround time in computer science.
Performance	It is minimizing or maximizing a parameter to be effective and efficient. Example: minimizing the latency of a job, maximizing resource utilization, etc.
Throughput	The amount of data processed or the number of tasks completed per unit time is called throughput.
Makespan	The time taken by a batch of jobs since its submission until its completion is called makespan.
Deadline	It denotes the finishing time of a job/task.
Budget	It represents the maximum money spent by a user to complete a job/task.

1.15 COMMON PITFALLS

1. *Are data mining and big data same?*
   No. big data refers to the data that is huge, heterogeneous, and high-speed, while data mining refers to the process of finding a hidden, unknown pattern from those data.

2. *Do you think that traditional* data mining and machine learning algorithms *can be applied to big data?*
   Of course, but traditional data mining and machine learning algorithms run on a single machine. Therefore, we need to construct a distributed and scalable algorithm that runs on many computers using tools like MapReduce, Spark, Storm, etc.

3. *Can we use incremental mining for handling big data in a single machine?*
   Certainly, great idea. However, big data cannot be maintained in a single server for a long time.

4. *What is the difference between P-P network and cluster computing? Which one is used for compute-intensive and data-intensive jobs?*
   P-P network is mainly used for data sharing/distributing services, not for data-intensive or compute-intensive tasks as there is no centralized management to coordinate nodes for executing tasks. So, we use cluster computing for data-intensive or compute-intensive tasks. The name 'cluster computing' became obsolete as we perform data-intensive or compute-intensive tasks on cloud computing. However, we use a cluster of VMs for Hadoop that run in a cluster of physical machines. Similarly, the name 'grid computing' also became obsolete. However, the hybrid cloud itself is grid computing.

KEYWORDS

- **big data**
- **big data framework**
- **big data systems**
- **distributed system**
- **Hadoop evolution**

CHAPTER 2

Hadoop Framework

The pessimist sees difficulty in every opportunity.
The optimist sees opportunity in every difficulty.

—Winston Churchill

INTRODUCTION

To write MapReduce job, we need to understand its execution flow and how data is framed in each step in the execution flow. Therefore, this chapters explains Hadoop distributed file system and MapReduce in detail and demonstrates with simple examples. Finally, we outline the shortcomings of MapReduce and some possible solutions to overcome.

2.1 TERMINOLOGY

It is essential to take a look at frequently used distributed computing terminologies before exploring more on Hadoop inner working concepts.

File system – is a software that controls how data is stored and retrieved from disks and other storage devices. It manages storage disks with different data access pattern like file-based (NTFS, ext4), object-based (swift, S3), block-based (cinder, database).

Distributed system – A group of networked heterogeneous/homogeneous computers that work together as a single unit to accomplish a task is called a distributed system. In simple words, a group of computers provides a single computer view. Example: Four computers each with dual core, 4 GB memory, and 1 TB storage in a cluster can be said as a distributed system with 8 cores, 16 GB memory, and 4 TB storage. The term "distributed" means that more than one computer is involved, where data/program can be moved from one computer to another computer.

Distributed computing – is managing a set of processes across a cluster of machines/ processors, and communicate/coordinate each other by exchanging messages to finish a task.

Distributed storage – A collection of storage devices from different computers in a cluster providing single disk view is called distributed storage.

Distributed File System (DFS) – Every file system software provides a namespace (path) to files in a tree structure. Local file system gives namespace only for the disks attached to it. DFS provides a unique global (unified) namespace on distributed storage. Example: if a user executes a command at node5 to display the value of "x" stored in node10, the result is displayed in node5 itself. User does not know where "x" is stored.

In short,

- Managing resources across a cluster of machines and providing a single computer view is called a distributed system.
- Managing a set of processes of an application across a cluster of machines or processors is called distributed computing.
- Managing a set of storage devices across a cluster of machines and providing a single disk view is called distributed storage.
- Providing a unique and unified global namespace on distributed storage is called DFS.

2.2   COMPONENTS OF HADOOP

Discussions made in this chapter are based on Hadoop 1.x in order to understand Hadoop 2.x better. Hadoop framework consists of a set of tools that work together for different purposes. There are two primary tools in Hadoop, upon which all other tools are installed.

Hadoop Distributed File System – storage part
MapReduce – processing part

Hadoop Distributed File System (HDFS) – When data outgrows the storage capacity of a machine, it should be partitioned and stored across a cluster of machines. HDFS is a software that allows us to store big data in a distributed environment across a cluster of low-cost, unreliable, commodity machines with streaming data access pattern. "Streaming data access pattern" means that the program reads and writes data from/to standard input and standard output. HDFS does not support update operation, so data in HDFS is Written Once and Read Many (WORM) times. HDFS has three sub-components: Name Node (NN), Secondary Name Node (SNN), Data Node (DN). HDFS is one of the file systems in Hadoop. There are other file systems: Local, WebHDFS, HAR, View, S3, Azure, Swift. URI scheme for these file systems is different as given below.

HDFS	– hdfs://IP:port#/path
Local file system	– fs:///path
Hadoop archive file	– har:///path

Although these file systems provide different features and possible to launch MapReduce jobs on all these file systems, it is always better to use HDFS while dealing with the huge amount of data.

MapReduce (MR) – MR is a distributed, scalable, fault-tolerant, data-parallel programming model for processing big data on a cluster of low-cost, unreliable commodity machines. MR allows writing distributed, scalable jobs with little effort, unlike programming with MPI. MR jobs can be launched in different file systems: HDFS, HBase, S3, local file system, etc. MR has two sub-components: Job Tracker (JT), Task Tracker (TT). MR is commonly preferred for the following applications.

- Searching, sorting, grouping.
- Simple statistics: counting, ranking.
- Complex statistics: PCA, covariance.
- Pre-processing huge data to apply machine learning algorithms.
- Classification: naïve bayes, random forest, regression.
- Clustering: k means, hierarchical, density, bi-clustering.
- Text processing, index building, graph creation and analysis, pattern recognition, collaborative filtering, sentiment analysis.

Figure 2.1 shows the components of HDFS and MR. These components can be launched in the same machine or different machines in the cluster as daemons. Daemon is a background process (like OS) running on the computer.

2.3   HADOOP CLUSTER

A data-center contains a set of racks stacked with a set of servers (nodes) forming a cluster with a high-speed local network. As Hadoop is based on shared-nothing architecture, each node has its storage HDDs. Therefore,

SAN/NAS/Direct Attached Storage (DAS) storage technologies are not used. Typical Hadoop setup in a physical cluster would seem as shown in Figure 2.2. Both HDFS and MR are based on master-slave architecture. NN, SNN, and JT are master/server components. DN and TT are slave components. In Figure 2.2, there are 4 racks each with 4 nodes. Server software components such as NN, SNN, and JT are run in separate nodes in different racks. Rest 13 nodes are called slave nodes running DN and TT together.

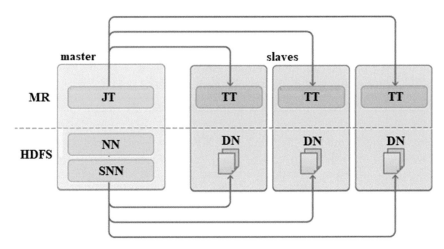

FIGURE 2.1   MR and HDFS components.

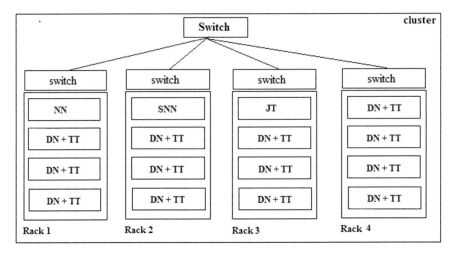

FIGURE 2.2   Hadoop cluster.

Figure 2.3 shows the communication between master and slave daemons. Master daemons talk to each other. JT communicates with TT, similarly NN talks with DN and SNN. TT talks to DN to bring data block for processing. In a Hadoop cluster, only one NN, SNN, JT, and multiple slave nodes can exist. Every slave node runs both DN and TT together. Master daemons (NN, JT, SNN) are usually run on a dedicated node to balance the load because NN and JT are profoundly communicated by slaves. Every slave node's resources (CPU and memory) are managed by TT for task execution. Storage disks are handled by DN for storing/retrieving data.

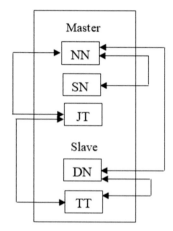

FIGURE 2.3   HDFS and MR components communication.

2.4   HADOOP DISTRIBUTED FILE SYSTEM (HDFS)

HDFS manages big data across a cluster of machines with streaming data access pattern. It employs distributed storage to provide a single disk view and DFS to provide unique global namespace on distributed storage. It is a specially designed file system with functionalities such as distribution, fault-tolerance, replication and deployed on low-cost, unreliable machines compared to other DFSs. HDFS is based on distributed and clustered file systems. Once HDFS is deployed on a cluster of nodes, storage disks of each node are logically pooled to appear as one large disk, as shown in Figure 2.4. From a user point of view, it seems to be a large centralized storage, but from a system point of view, it is the individual server that is contributing its storage.

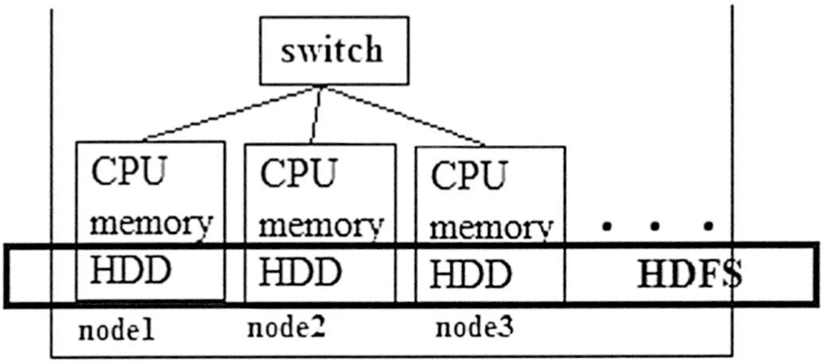

FIGURE 2.4   Distributed storage.

HDFS provides file abstraction, which means that a file beyond a storage disk size is partitioned into chunks and stored across a cluster of nodes. For a regular user, the huge file is logically shown as a single file, but in reality, parts of this file are stored in different nodes in the cluster. File abstraction can be achieved using NFS as well. However, HDFS uses a local file system to implement DFS compared to NFS. HDFS is immutable, meaning that only the initial load is possible. There is no file editing facility in HDFS using vi, gedit, etc. as it involves huge IO overhead.

Use HDFS when you want

- to store large data (over TB) on commodity servers.
- to process a small number of large files than a large number of small files.
- batch reads instead of random reads/writes.

Do not use HDFS

- for transaction process, and low latency access (in ms).
- to process a lot of small files (HDFS takes more meta-data, and resources).
- for parallel write (HDFS supports only append mode).
- to arbitrary read (HDFS provides only batch read).

HDFS provides file permissions and authentication. HDFS can extend its storage capacity from the non-HDFS portion of local storage. Once data is loaded into HDFS, you can run MR jobs, hive/pig queries. Let us understand each component of HDFS in detail.

2.4.1 NAME NODE (NN)

NN is a centralized service/daemon that acts as a cluster storage manager. Its primary responsibilities are:

*   maintaining meta-data of files and directories.
*   controlling and coordinating DNs for file system operations such as create, read, write, etc.

Clients communicate with NN in order to perform everyday file system operations. In turn, NN gives clients the location of DNs in the cluster (where the data block is available) to carry out operations. NN does authentication if it is configured.

2.4.1.1 FILE BLOCKS

The unit of data storage and access in HDFS is a block, which denotes the minimum amount of data that can be stored and retrieved from HDFS. Data we wanted to load onto HDFS is divided into equal sized chunks (called blocks) and stored across DNs in the cluster. DNs store each data block as a file on its local file system. As a result, access to an original file usually requires access to multiple DNs to reassemble the blocks into a file. To understand blocks, we need to understand the internal structure of the HDD. An HDD is composed of a set of platters (disks). Each disk is designed with many tracks, which comprise a set of sectors (physical blocks) of size 512 B. Physical sector size denotes a minimum amount of data to read/write from disks. Linux File system uses a logical block size of 4 KB (a group of multiple physical blocks) as shown in Figure 2.5.

Small blocks are suitable for transactional databases. Large blocks are useful for analytical databases. Linux file system maintains meta-data at the file-level. HDFS uses the logical block size of 64 MB. So, it maintains meta-data at the block-level. HDFS is a logical file system on top of the local file system, as shown in Figure 2.6. It is not a real file system like ext3, ext4 that works with physical disk-level. So, HDFS runs on local file systems (ext3/ext4) and does not directly interact with storage devices.

A file which is bigger than logical block size is broken up into multiple blocks. Blocks are indexed, but not the contents inside the blocks. By default, the block size is 64 MB in HDFSv1. However, you can use a block size of

128/256 MB if dataset size is over PB. A file which is smaller than a block size is considered to be one block but does not occupy a full logical block size. For instance, if the file size is 10 MB, only 10 MB is occupied, not 64 MB, unlike Linux file system that occupies the entire 4 KB even if the file size is 1 KB. In this case, arbitrary file size (<64 MB) itself is considered to be a block size. Every file we upload onto HDFS should have at least 1 block. Directories will not get any blocks.

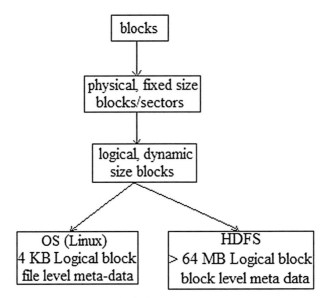

FIGURE 2.5   Physical vs Logical block size.

HDFS block size is large when compared to disk physical block size. The reason is to minimize the cost of seeks in disks than block transfer rate. By making a logical block size large enough, seek time is reduced than the data transfer rate. To achieve this, seek time must be 1% of the data transfer rate. Example: If the data transfer rate is 100 MB/s, we can transfer a complete block (64 MB) per second. Therefore, seek time is supposed to be 10 ms. Thus, blocks are transferred faster without keeping disk bandwidth idle.

2.4.1.2   WHY SHOULD BLOCKS BE THE EQUAL SIZE?

Consider 1 GB file to process. If this file is loaded onto HDFS with different block size (four 100 MB blocks, five 50 MB blocks, five 20 MB blocks),

19 map tasks will be launched to process these 19 blocks. Notable points here are:

- Job latency is determined by the map task processing large block.
- The time required for scheduling and spawning tasks for small blocks dominate its processing time.

Therefore, it is essential for blocks to be of the same size to carry task execution and its resource allocation uniformly. Example: If there is a file of 200 MB and HDFS block size is 64 MB, there will be three 64 MB blocks and one 8 MB block. Mostly, the last block of a file may not always be the block size.

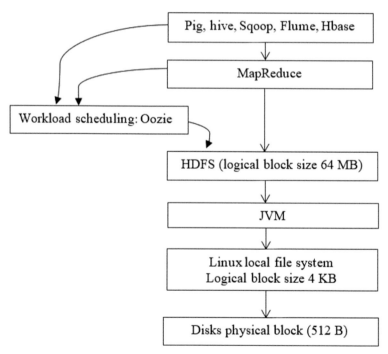

FIGURE 2.6 HDFS on local file system.

2.4.1.3 META-DATA

One of the primary responsibilities of NN is to maintain meta-data of files and directories. Meta-data typically comprises three parts in memory: FSImage (read-only) file, edit_log file, and block location information. FSImage and edit_log files are persisted (written in disk) and maintained in NN memory as well. However, block location information is maintained only in memory. In Figure 2.7, there are four different datasets/files (denoted in different colours), which are divided into blocks and stored in DNs. Meta-data of these datasets and its respective blocks are stored persistently in two different files in NN disk: FSImage and edit_logs. Attempting to modify meta-data will cause HDFS downtime and consequently, permanent data loss.

FSImage – File System Image is a read-only file that contains all sorts of information about input files and blocks such as file name, file ID, list of blocks of a file with its ID, namespace (path), ownership, permissions, storage quotas, Replication Factor (RF), date of creation of directories/files/blocks, etc., as shown in Figure 2.7. Here, namespace is just a path to an object (file/block) in HDFS. NN manages file system namespace (path) in a tree structure.

Edit_logs – During file system operations such as read, block attributes are referred in FSImage. For file system operations such as write/delete, FSImage is not updated as it is read-only. Because, updating FSImage file for every file system operation now and then is computationally expensive. Moreover, until an FSImage update is finished, it is not available for any read operation. Therefore, little downtime is inevitable. To avoid this, after initial construction of FSImage file, every write/delete operation information is maintained in edit_log files separately in memory. These edit_log files are periodically persisted in NN disk. For every write/delete operation, a unique monotonically increasing transaction ID is assigned. Over period of time, many edit_log files are created and referred for read operation after FSImage file because latest block information is available edit_log files only. At this moment, FSImage file represents the complete file system state up to a specific transaction ID. Edit_log contains the transaction records of current modifications made after the most recent FSImage. These edit_logs are also called journals. More detail on meta-data generated by different components is available in Section 6.21.

Block location information – When billions of blocks are stored in HDFS, referring FSImage and edit_log files for read/write operation is time consuming. Therefore, block location of every file is maintained separately in memory for faster access. Using this, NN performs block mapping (which block belongs to which file and stored in which DN). This block information is not persisted and lost if NN is down. When NN is up, it collects block location information from all DNs via block reports. Until this process is complete, no block write operation can be performed as NN is on safe-mode, in which only read operation is permmitted.

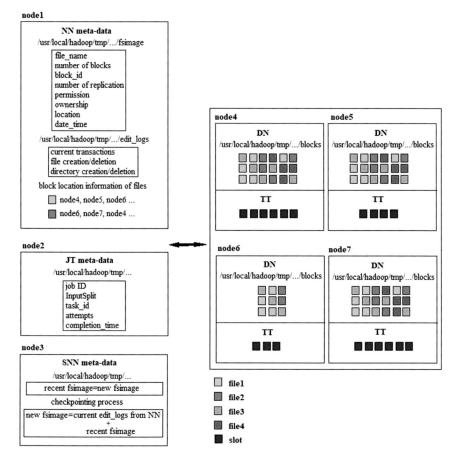

FIGURE 2.7 HDFS and MR meta-data.

Meta-data for objects (file, block, and directory) take 150 bytes by default. A typical meta-data of files/blocks in HDFS would be shown as below.

$ hadoop fs -ls /   // issue this command after Hadoop setup

Column: **1**   **2**   **3**   **4**   **5**   **6**   **7**   **8**
drwxr-xr-x 1 itadmin supergroup 0 2017-05-30 21:34 /input
-rw-r--r-- 3 itadmin supergroup 23170 2017-05-31 14:54 /largedeck.txt

Column 1: denotes the file mode (access permission). There are ten letters which are read from left to right. The first letter denotes the object type: file or directory. If it is directory, letter d is set. If the uploaded object is a file, hyphen symbol (–) is used. The next three characters specify permission for the user, and the next subsequent three characters specify the access permission of users in the same group, and the last three letters signify the access permission of non-group users. Three characters denote the access permission: read (r), write (w), executable (x). If anything is not applicable, hyphen (–) symbol is used.

> **r** – read permission is required to read files or lists the contents of a directory in HDFS.
>
> **w** – write permission is required to write a file/directory in HDFS.
>
> **x** – denotes executable, and it is not for files, because you cannot execute a file in HDFS. This permission is required to access the children of directories.

By default, there is no authentication mechanism enabled in NN to check for the owner, group user, non-group user, etc. to permit file access.

Column 2: indicates RF, which is not applicable for directories. Because directories are not replicated and treated as meta-data stored in NN. DNs store only blocks (as local files). There is nothing stored about directories in DNs.

> Column 3: username (owner).
>
> Column 4: group name.
>
> Column 5: file size in bytes, which is zero for directories.
>
> Column 6 and 7: last modified date and time.
>
> Column 8: file/directory name.

Example: Consider to store 1TB of data on HDFS. Calculate the amount of memory required for meta-data when you use the local file system and HDFS.

Local File System: logical block size is 4 KB, by default.

> = ((1TB/4 KB) x 150 B) = 37.5 GB memory is required to maintain meta-data for all the objects. Managing this huge meta-data in memory is impossible for a faster response as HDFS does not support demand paging.

HDFS: logical block size is 64 MB, by default.

> = ((1TB/64 MB) x 150 B) = 2.34 MB memory is enough to maintain meta-data of all the objects. It is easy to keep in memory for faster access.

How about block size of 128 MB or 256 MB in HDFS? Meta-data takes very less memory. That is the reason we use block size greater than 64 MB. So, indexing takes very less effort. If there is a large number of small files, each file will have at least one block. Meta-data for all these objects will increase. Thus, NN will take more time to find the block location, sometimes, NN may not be able to maintain the whole meta-data in memory and there is no concept of demand paging in HDFS.

2.4.1.4 REPLICATION PLACEMENT USING RACK AWARENESS

As Hadoop deployment is mostly done with unreliable commodity servers, the failure of nodes is more likely. Therefore, computation and data loss are inevitable. How does Hadoop take care of block loss? HDFS does not rely on RAID. Because RAID only solves HDD failure and mirroring is expensive for big data. Consider an input file having three blocks A, B, C, and 12 nodes in the cluster, as shown in Figure 2.8. If you store all three blocks in node1, and if the node1 is crashed or disk failed, it is not possible to get the original file back. Therefore, HDFS ensures fault-tolerance by replicating blocks to different nodes by itself. The number of replications of data blocks is denoted as RF, which is three by default. Replication helps to achieve fault-tolerance, network performance, and improved data-parallel processing. Please note that only blocks of a file are replicated, not directories. We can run the same task of a job on block A, B, C at the same time, so that data parallelism is achieved. However, the same task will not be executed in two copies of block A simultaneously. On the other hand, tasks of two different jobs can run on two copies of block A.

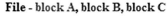

File - block A, block B, block C

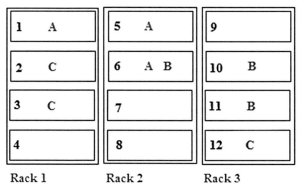

FIGURE 2.8 Replica placements in HDFS.

If you store three copies of a block in the same machine, if the machine is down, none of the copies will be accessible. So, that particular block is lost. Therefore, there is no point in making more than one copy of a block in the same machine. It wastes memory and provides no fault-tolerance as well. So, each copy is distributed to more than one node in the cluster. The rule of thumb is that RF must be less than or equal to the number of nodes in the cluster. NN decides where to place the replicated blocks based on cluster topology, as shown in Figure 2.8. This is called as rack awareness. Each block is replicated with three copies. For instance,

- The first copy of block A is stored in node5 of rack2.
- The second copy of block A is stored in the same rack, but in a different node (node6).
- The third copy is stored in node residing other than rack2.

It is called a rack/topology-aware data block placement. However, why is such a placement scheme required? Consider the following cases, as shown in Figure 2.9:

Case 1: What if node3 failed or data block is corrupted?
Case 2: What if the rack1 is down due to network switch failure?

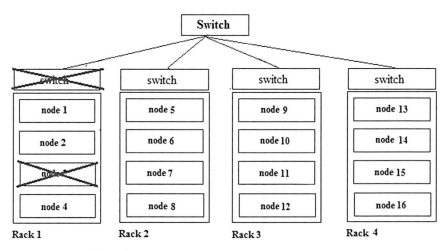

FIGURE 2.9   Failure of DN and network switch.

How can we reconstruct the original file if these two cases happened?

Case 1: HDFS replication solves node failure, disk failure, and block corruption. As shown in Figure 2.10, if node3 in rack1 crashed or disk failed or block corrupted, a copy of same block C is available in node2 of rack1 to reconstruct the original file.

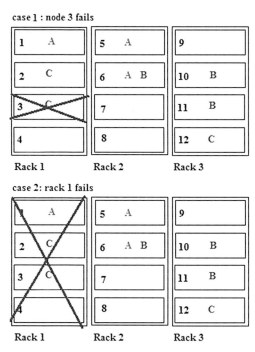

FIGURE 2.10 File reconstruction in case of DN and network switch failure.

Case 2: HDFS replication overcomes network failure (rack failure) problem for data block. If you have all three copies in different machines of the same rack, none will be accessible if the top of the rack switch failed. Therefore, by default, the first two copies are stored in two different machines of the same rack, and the third copy is stored in the node of some other rack. Thus, rack awareness helps to avoid the loss of data when the entire rack is down.

What if multiple top of the rack switches fails together? Well, this cannot be solved by HDFS. Data center engineer may have to take serious actions. By default, all nodes are added to default rack with the flat network. To enable rack awareness, we have to manually feed the cluster topology file to HDFS. Because, Hadoop does not automatically discover cluster topology.

Bandwidth is a scarce resource in data-center. Intra-rack bandwidth is higher than inter-rack (cross rack) communication. So, to minimize the data block transfer across cluster frequently, distance is calculated based on server location and used by NN during block placement. MR scheduler also uses the distance to determine where the closest replica is available for launching map/reduce tasks. Typical distance of servers in a cluster is given below.

D = 0 same node
D = 2 distance between nodes in the same rack
D = 4 distance between nodes in different racks
D = 6 distance between nodes of different data-center

2.4.2   SECONDARY NAME NODE (SNN)

NN maintains the meta-data of entire HDFS cluster. If NN is down, the entire HDFS cluster becomes inaccessible because of meta-data unavailability. This is called Single Point Of Failure (SPOF). Therefore, it is recommended to run NN in an enterprise-grade server in a production environment. However, the failure of servers is inevitable. Therefore, a copy of meta-data from NN is maintained in another machine, called SNN. An alternative option is using NFS to maintain meta-data.

As we know, FSImage is only referred, not updated. Therefore, latest operations are recorded in edit_log files. When edit_log file size goes beyond a threshold, referring edit_logs for read operation takes more time. Therefore, edit_log files are sent to SNN. SNN receives current edit logs from NN and merges (as shown in Figure 2.7) with the previous FSImage (got from NN) to form new FSImage. This process is called checkpointing. This process requires enough memory and CPU to perform faster. That is the reason why SNN is run in separate machine. After merging, SNN sends the new copy of FSImage file to NN. Once this is done, old edit_logs in NN memory are deleted up to the last checkpointed transaction ID.

Checkpointing process can be triggered automatically (if time is expired or threshold size is reached or up to a particular number of operations performed) by configuration policies or manually triggered by HDFS administrative commands. The fs.checkpoint.period property is used to set the interval between two consecutive checkpoints. By default, the interval time is in 3600 seconds (1 hour). Edit log file size is specified by parameter fs.checkpoint.size (default edit logs size 67108864 bytes, that is 64 MB in memory). Multiple checkpoint nodes may be specified in the cluster. SNN does not provide High Availability (HA) (also called hot-standby) or auto-failover. That is, If NN fails, SNN will not become as NN automatically. A new node is introduced into the cluster (SNN is preferred), which restores FSImage from SNN and starts serving as NN. All these activities involve manual effort. During NN restart, the major steps done in NN are:

- NN restarts in safe-mode (does not provide write access to DNs) until all block location information is obtained.
- Restores latest FSImage from SNN.
- Gets block reports from all DNs to determine the block locations.

It takes 5 to 15 minutes to bring new NN up and stable in the large cluster as millions of blocks stored. If NN failed before sending the current edit logs to SNN, SNN lags the current state to produce new FSImage. Therefore, the loss of information is almost inevitable. In order to avoid this problem, HDFS v2 has introduced NN HA.

2.4.3   DATA NODE (DN)

DN daemon is responsible for managing local storage disks. It handles file system operations such as creating, reading, opening, closing, deleting blocks. The DN does not know anything about the data blocks. It stores each block as a file in the local file system. The DN does not store all blocks in the same directory. Instead, it uses a heuristic to determine the optimal number of blocks per directory and creates sub-directories appropriately. Blocks are loaded/ deleted in DNs based on the instructions of NN, which validates and processes requests from clients. NN does not perform any read/write operations for clients. Clients communicate with NN to know the locations of the blocks and redirected to DNs to perform read/write operation.

2.4.3.1   WHO SPLITS A FILE INTO BLOCKS?

Consider two racks each with four nodes running DN+TT and four separate nodes to run HDFS client, JT, NN, SNN in the cluster. A user remotely connects to HDFS client to upload big data onto HDFS. Figure 2.11 illustrates the flow of data block read/write. HDFS client is responsible for dividing an input file into blocks and forwards the user request to NN. This daemon can run in a separate node in the cluster or co-locate with other HDFS services (NN/DN/SNN). A node which runs HDFS client is also called a gateway node.

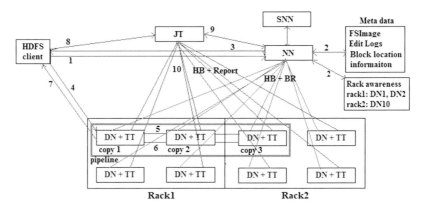

FIGURE 2.11   Block read/write in HDFS.

2.4.3.2   WRITING FILES ONTO HDFS

Step 1: Once HDFS client receives the file upload request, it divides the file into blocks physically and requests NN where to store all these blocks. Because HDFS client does not know which DN has free space and where to store based on rack awareness.

Step 2: NN goes for various checks: file already exists, user access permission, etc. It would return an IOException if these checks failed. NN refers to meta-data (FSImage file) and determines three locations based on rack awareness for each block. Then, NN sends a list of three DNs locations for each block. Say, locations for block1 are DN1, DN2, and DN3. Suggesting the number of DNs depend upon the RF. By default, three DNs are suggested for each block as default RF is 3. If the client node is in the Hadoop cluster, it is the first node to copy the data block.

Step 3: Upon receiving three locations for each block, HDFS client is ready to write blocks into respective DNs in parallel by opening HDFS output stream. There is no constraint that block 2 should be copied onto HDFS only after block1. All blocks are copied to the cluster depending upon the number of threads an HDFS client can serve.

Step 4: HDFS Client further logically divides block1 into a queue of packets (each 64 KB), called data queue. For block1, HDFS client forms a pipeline with DN1, DN2, and DN3.

Step 5: First packet from the data queue is copied to DN1, which stores and forwards that packet to DN2 along the pipeline.

Step 6: There is an acknowledgment queue in HDFS client, which has to confirm the acknowledgment of each packet sent to the data pipeline. For each packet, DN3 sends Ack to DN2, and DN2 sends Ack to DN1, which in turn sends Ack to HDFS Client. DN1 keeps receiving packets from HDFS Client regardless of DN1 forwarding packets to DN2.

What happens if there is an error in the pipeline?

Assume DN2 has no space or network failure, and DN1 and DN3 are good. By default, all DNs update NN about the data blocks stored via Block Report (BR). DNs send Heartbeat (HB) to NN to indicate it is alive and ready to serve. If DN2 does not send any HB to NN due to network failure or space unavailability, NN removes DN2 from the pipeline. Now, data pipeline comprises only DN1 and DN3. However, data loading continues from where it was left until all blocks are loaded in DN1 and DN3. Finally, DN1 and DN3 send BR to NN to update the locations of new blocks. Now, NN observes the number of copies for block1 and finds one copy of block1 is missing (under replication). So, NN instructs either DN1 or DN3 to copy the block1 to some other DN, say to DN8, which has free space. Again, a data pipeline is constructed with only DN8. NN keeps updating its metadata using BR of DNs. Importantly, DNs communicate with each other only during block replication.

It is very less likely to happen that all DNs fail in the data pipeline. However, for a block write to complete, at least, one copy should be written into HDFS, else block write is failed. Then, the HDFS Client will get a new set of DNs from NN to form a data pipeline. Packet replication is done sequentially and synchronously. However, multiple blocks can be moved simultaneously onto HDFS by creating its pipeline.

Step 7: HDFS Client receives an acknowledgment from DNs stating that the block has been copied successfully. HDFS Client sends a success message to the user. All these processes are highly transparent to the user. User does not know where data blocks are stored. MR job can be launched only after all blocks are successfully loaded into HDFS. The user also has a privilege to specify block size, RF while uploading big data.

2.4.3.3   READING FILES FROM HDFS

Step 1: User launches a read command to HDFS Client, which forwards the request to NN to find the location of blocks for the requested file.

Step 2: NN refers to its meta-data and finds a list of DNs where each block has been stored. This list is prepared based on the location (rack awareness) of block copies.

Step 3: HDFS Client is ready to read blocks from different DNs by creating HDFS input stream.

Step 4: HDFS Client prefers DNs that are near to the client (to reduce network traffic). If the first DN in the list is not reachable, the second DN in the list is chosen.

Step 5: Blocks of a file are read sequentially one by one according to the order to construct an original file. There is no use of reading multiple blocks simultaneously. Finally, DNs stream data blocks to the client in order.

2.4.3.4   HEARTBEAT (HB)

HB is very similar to ping command and very essential for a scalable architecture. NN periodically (dfs.heartbeat.interval – default 3 seconds) receives HB from each DN in the cluster to ensure whether a DN is alive and ready to serve. If DN does not send HB for 10 minutes, the particular DN will be considered dead and removed from HDFS cluster by NN. Blocks stored in that DN are replicated to another DN in the cluster. HB from DN to NN carries the following information:

- Registration: DN registration information.
- Capacity: available storage capacity in DN.
- HDFS-Used: storage used by HDFS.
- Non-HDFS: remaining storage available with other file systems.
- blockPoolUsed: storage used by the block pool.
- cacheCapacity: total cache capacity available in DN.
- cacheUsed: the amount of cache used.

This information is used by NN in the following ways:

- The health of DN: should this DN be marked dead or alive?
- Registration of new DN: if this is a newly added DN, its information is registered.
- Update the metrics of DN.

In general, NN does not directly instruct DNs. It just responds HB to instruct DN. Some typical instructions are:

- replicate blocks to other DNs.
- remove local block replica.
- re-register or shutdown the DN.
- request DN to send an immediate BR.

2.4.3.5   BLOCK REPORT (BR)

BR is a message comprising meta-data of all blocks present in a DN. It is sent periodically by DNs to NN to update the blocks available in HDFS cluster. NN updates the meta-data of blocks upon receiving BR from DNs. Whenever DNs boot, they scan through the blocks stored at present and create a BR to send NN.

The first BR is sent immediately after the DN registration. Subsequent BR is sent periodically (by default, dfs.blockreport.intervalMsec – 21,600,000 ms (6 hours)). Some of the information contained in BR is block ID, block size, and block generation timestamp. NN always analyzes under/over replication of blocks upon receiving BR. Under replicated blocks are put into priority queue to balance RF with high priority. NN will mark the over-replicated blocks to remove from the HDFS.

2.4.3.6   DATA INTEGRITY

When we store huge data in the cluster of commodity servers, and access through network devices, chances of data loss/corruption are high. However, HDFS assures that no data will be lost while storing/transferring. To achieve data integrity, HDFS calculates checksum while storing data blocks. Later, when data is read from disks, the checksum is again calculated and verified against the checksum calculated during initial loading. If it did not match, the corrupted data block is informed to NN, which will not direct any accesses to the corrupted blocks. Then, NN replaces the corrupted data block with a fresh copy of the same block to maintain RF. The checksum is merely an error detection technique and does not correct the corrupted data.

Possibility of checksum data corruption is very less, as it is very smaller. Hadoop uses CRC-32C technique to calculate the checksum. The checksum is calculated for every 512 bytes (by default dfs.bytes-per-checksum) of data when HDFS Client sends blocks through the data pipeline. CRC-32C is 4 bytes in length, which ensures storage overhead less than 1%. Last DN in the data pipeline verifies the checksum and sends IOException if there is any corruption. Then, those particular 512 bytes are resent by HDFS client. DN verifies checksum every time it reads from disks and maintains the details of verification in checksum log. Logging such information helps in detecting bad disks.

2.4.3.7   DATA BLOCK SCANNER

Blocks are stored in DNs along with meta-data such as checksum to verify data integrity. Each DN has a thread "DataBlockScanner" that verifies block checksum periodically and update NN. Corrupted blocks will be informed to NN, which will replace it with a fresh copy taking from some other DN. This allows bad blocks to be detected and fixed before clients read them. Blocks are verified every three weeks (dfs.datanode.scan.period.hours defaults to 504 hours) to guard against disk errors.

2.4.3.8   BLOCK BALANCER

To increase cluster storage and computation capacity, we add more slave nodes. This results in cluster imbalance as new nodes initially do not have any data stored. Therefore, in order to improve data-parallel processing, every DN in the cluster should have some data blocks stored. To achieve data load balancing among DNs, we use a command to redistribute blocks from overloaded DN to empty DN by satisfying RF and topology-aware policy. This takes several minutes to distribute data blocks across DNs evenly. Rebalancing process continues until cluster load is balanced or until all the options are tried out to move blocks. The balancer is designed to run in the background without disturbing the running daemons. It uses minimal local bandwidth, 1 MB/s (dfs.datanode.balance.bandwidthPerSec in bytes), to copy from one node to another because it should not disturb the running jobs.

2.4.3.9   BLOCK CACHING

Frequently accessed blocks can be cached in DN memory itself. Schedulers can use these blocks for map tasks instead of loading them again from disks.

This improves read performance and minimizes latency. Users also can specify which data blocks of a file to be cached. This is highly useful for join operation.

2.5 MAPREDUCE (MR)

MR is highly distributed, horizontally scalable, fault-tolerant, data-parallel batch processing tool in Hadoop framework that runs on a cluster of unreliable commodity servers to process big data. Data is collected and stored onto HDFS before launching MR jobs. Programming with MPI for distributed parallel processing is limited in scalability, moreover, programmer responsibilities are huge as mentioned before. In contrast, it is easy to develop scalable algorithms using MR with no extra effort to manage a distributed system. A program developed for one node can be used for 1000s of nodes without re-modifying the code. MR itself parallelizes the execution, so users need not invest effort for parallel execution.

MR programming is based on LISP (LISt Programming), a type of the functional programming paradigm (like Haskell, Scala, Clojure, Smalltalk, ruby). Everything in functional programming revolves around function. A function can be passed/received as arguments, distributed across nodes. A function taking another function as an argument is called a higher order function. Functional programming does not maintain state and does not support blocking and synchronization. Hence, it is scalable. MR tasks run independently, so it is easy to handle partial failure.

All applications are not suitable to implement with MR. Only applications that need functions like counting, ranking, aggregation, searching, sorting, and grouping can be accomplished with MR. MR does not have any predefined queries like OLAP for DWH. It is all about writing ad-hoc algorithms to process the entire dataset. MR performance is lesser than general programming languages when processing datasets of few 100 MBs, because MR consists of a sequence of steps to be carried out. However, to experience the real power of MR, one should work with dataset in TBs, because this is where RDBMS takes hours and fails, whereas Hadoop does the same in minutes. As discussed already, MR has two sub-components (see Figure 2.12): JT and TT. A typical MR job executes two tasks: map and reduce.

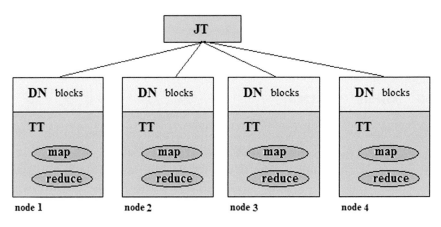

FIGURE 2.12   MapReduce components.

2.5.1   JOB TRACKER (JT)

JT is a daemon running in a dedicated server in the cluster to manage cluster resources and MR jobs. The primary responsibilities of JT are

- managing and monitoring resources (CPU and memory) in TTs.
- preparing execution plan and phase coordination.
- scheduling map/reduce tasks to TT.
- jobs life cycle management (from the time of submission till completion).
- maintaining job history (job-level statistics).
- providing fault-tolerance.

JT is also rack-aware while scheduling map/reduce tasks. Most of the time, processing takes place at nodes where the required data block is physically available to reduce network traffic. It is called data locality/gravity. JT also suffers from SPOF.

2.5.2   TASK TRACKER (TT)

TT is a daemon that runs in every slave node in the Hadoop cluster. It manages the local resources (CPU and memory) as a slot. A slot (also called as container) is a fixed logical pack of a portion of memory and CPU to run map/reduce task, as shown in Figure 2.13.

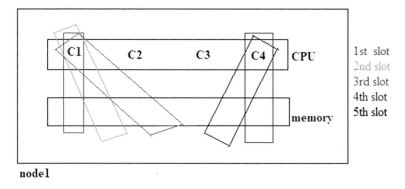

FIGURE 2.13 Slot.

A CPU/processor in a node may contain more than one physical cores or logical cores (C_1, C_2, etc.). Please refer Section 5.5.2 for more information about logical cores. By default, a slot is formed with 1 logical core and 1 GB memory. There can be many slots possible in a TT, and the number of slots is determined based on the available CPU and memory size. In MRv1, the number of map and reduce slots in a TT is fixed. For instance, if there are four slots in a TT, we may assign three slots for map task execution and one slot for reduce task execution. This can be adjusted depending on our requirements. However, a necessary restriction is, map task cannot be assigned to reduce slot and reduce task cannot be assigned to map slot. JT monitors the availability of slots to assign appropriate tasks in TT.

TT is responsible for starting and managing individual map/reduce tasks in its dedicated slots. TT periodically sends a report to JT about resource availability, task progress, and its status. TT sends HB to JT to indicate ready for service. As you add more nodes in the Hadoop cluster, the number of task slots increases linearly, and more data blocks can be processed in parallel. Therefore, the number of blocks processed per unit of time (throughput) increases linearly.

From Figure 2.11, Step 8 to Step 10 are as follows:

Step 8: Assume data is already uploaded onto HDFS. Now, an MR job is submitted to Job Client, which is a daemon that is part of JT.

Step 9: JT will query NN about the block locations of a specified input file. NN responds with a list of three DNs having copies of each block based on rack awareness.

Step 10: JT prepares an execution plan and performs phase coordination. Similar to NN, JT also is highly rack/topology-aware. It chooses the nearest DN having desired block at first. If there is no free slot in the TT, then next DN is chosen to achieve data-local execution. If there is no free slot in the given three slaves which have the desired copy, a copy of that block is taken to any other TT which has a free slot to execute. This is called non-local execution, which consumes more network bandwidth. Non-local execution is preferred only when there is no possibility of achieving data-local execution for some amount of time.

It is always good to launch MR job after the complete file is loaded into HDFS because it is not guaranteed that all blocks of a file have been written into HDFS.

2.5.3   MAP AND REDUCE TASK

To develop MR job for an algorithm, one must write map and reduce functions. Some parts of an algorithm are included in map function, and some are included in reduce function. To achieve this, we should have a strong understanding in MR execution sequence (refer Section 2.6) to decide which part of the algorithm should be included in the map and reduce functions. One can implement logic such as filtering, projection, transformation in the map function. Similarly, logic such as aggregation, join, sorting can be implemented in reduce function. When a function takes input and produces output, it is called as task. As in Figure 2.20, when a map or reduce function takes input and produces output, they are called tasks: map task and reduce task. These tasks are executed in dedicated slots in TT. In general, map task is called as mapper, and reduce task is called as reducer. A node that runs map task is called a map node, and a node that runs reduce task is called a reduce node. Map and reduce functions read/write data from standard input/output. This facilitates any language (Java, C++, Ruby, Python, etc.) that supports standard input/output is allowed to write MR programs.

A typical MR job consists of a set of map tasks and a set of reduce tasks. When a job is launched, first all map tasks are executed, then all reduce tasks are executed. Map tasks do not have dependency among them; similarly, reduce tasks also do not have dependency among them. Therefore, these tasks are called "bag of tasks," which support huge parallelism with concurrent execution. This is also called embarrassingly (or perfectly or pleasingly) parallel execution. MR is highly asynchronous as there is no communication, blocking, and synchronization among map tasks and reduce tasks. However, there exists a synchronization point, where reduce tasks will be started only after all map tasks completed. So, map tasks are given high-priority than reduce tasks as it has to complete before reducer can start. Map tasks are scheduled to achieve data locality to the extent. There can be non-local executions too. Reduce tasks do not have the advantage of data locality. JT dynamically decides where to run reduce tasks based on the slot availability and network load.

2.6 MAPREDUCE PHASES

In general, there are two phases for a MR job execution, as shown in Figure 2.14: map phase and reduce phase.

- Map phase executes a set of map tasks (mappers) to read data from disks as key-value pairs and produce an arbitrary number of intermediate key-value pairs based on user-defined map function.
- Reduce phase executes a set of reduce tasks (reducers), which collect the output from all map tasks, merges, sorts based on key, groups list of values that belong to the same key and produces the final output based on user-defined reduce function.

For example, in map phase, two map tasks are running in node1, and one map task is running in node2 and node3. In reduce phase, one reduce task is running in node4 and node5. All map and reduce tasks are executed in any node in the cluster. Map phase and reduce phase contain a sequence of steps, as shown in Figure 2.15, to be carried out for data processing. Every step in the MR phase takes input and produces output in the form of key-value pairs.

HDFS   : Input file → Blocks → Map phase

Map phase   : FileInputFormat → Input Split → Record Reader → Mapper → Partitioner → Combiner

Reduce phase : Shuffle → Merge → Sort → Group → Reducer → FileOutputFormat → Record Writer → HDFS

HDFS   : Blocks → Output File

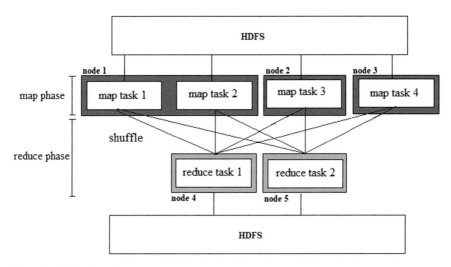

FIGURE 2.14   MR phases.

2.6.1 MAP PHASE

HDFS physically divides the submitted input file into equal-sized blocks and stores in different DNs based on rack awareness. Required data blocks are brought into memory to feed the map task for execution to begin map phase. Kindly refer Section 2.7 and Section 2.8 while reading further for better understanding. Map phase begins from FileInputFormat and ends when all map tasks completed. Map node for a map task executes the following sequence of functions to be executed along with map function: Input Split → Record Reader → Mapper → Partitioner → Combiner

Step 1: FileInputFormat and Input Split (IS) or Splits

An IS represents a set of blocks to be processed by an individual map task. Block is HDFS concept whereas IS is MR concept. IS is the logical grouping of one or more physical blocks. IS will refer to at least one block. By default, IS size is the same as the default block size (64 MB). As shown in Figure 2.16, 256 MB input file is divided into four 64 MB physical blocks and stored in different DNs. If IS size is 128 MB, then each IS logically groups two physical blocks. Each IS is processed by a map task. Please note that IS does not contain a copy of physical blocks. It just contains the location of physical blocks and its meta-data.

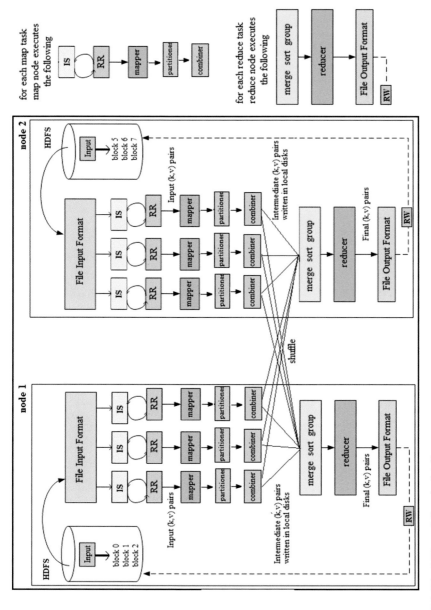

FIGURE 2.15 MR workflow in detail.

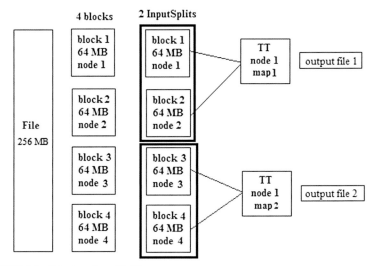

FIGURE 2.16   Block vs InputSplit.

A map task can process only one IS in a given time. The number of map tasks for a job is driven by the number of IS. The number of map tasks launched is equal to the number of IS formed. On the other hand, it can be customized by setting the block size. As shown in Figure 2.17, assume an input file size of 1 GB with varying block size (64 MB, 128 MB, 256 MB). Calculate the number of map tasks launched for these different block size.

As we know, by default, IS size is the same as the block size. With this hint, we can calculate the number of IS when physical block size changes; thereby, we can count the number of map tasks launched.

Number of IS = input file size / HDFS block size = number of map tasks

= 1 GB / 64 MB = 16 map tasks
= 1 GB / 128 MB = 8 map tasks
= 1 GB / 256 MB = 4 map tasks

Similarly, by changing IS size, we can adjust the number of map tasks. For instance, consider setting block size 64 MB and IS size 128 MB. Now, each IS logically packs two blocks. If input file size is 1 GB, the number of IS is (1GB/128MB) 8. Therefore, 8 map tasks are launched each processing two blocks. As you increase IS size, the number of map tasks is minimized. However, the latency of a map task increases due to a greater number of data blocks to process as there is no scope for parallelism.

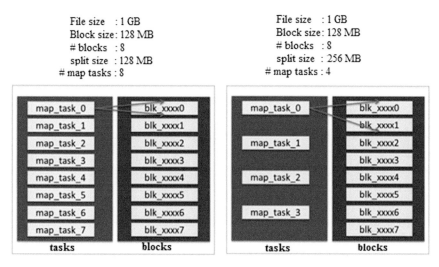

FIGURE 2.17 Calculating number of map tasks.

What is the IS size?

By default, block size itself is IS size. However, in order to customize IS size, set minimum and maximum IS size to be the same. We can set the size of IS by using properties mapred.min.split.size and mapred.max.split.size. If the block size is 64 MB and mapred.min.split.size==mapred.max.split.size=256 MB, then each IS logically groups and points to four physical blocks. So, only four map tasks will be launched. This can be done per job basis. However, for performance reason, the recommended IS size is the block size itself. All IS is ordered based on size. So, the largest IS is executed first. The size of IS can be calculated as below.

IS size= max(minimum IS size, min(maximum IS size, blockSize))

So far, we discussed how the number of map tasks is determined by modifying the block size and IS size. Setting block size can be done only at the time of data loading. IS size can be specified whenever a MR job is launched. It is also possible to set the number of map tasks, but it is not recommended.

What if an IS contains more than one data block?

If an IS contains more than one data block, data locality is sacrificed. Consider an IS that contains three data blocks (A, B, C) residing in different nodes: A in node1, B in node2, C in node3. A map task is launched in any one of the

TTs containing relevant block. Say, map task is launched in node3 to process block C. Now, other two blocks (A and B) from node1 and node2 are copied to node3. This results in high latency as blocks are copied over the network to achieve non-local execution and introduce more network traffic. This is the drawback when we set IS size beyond the block size. Therefore, it is always recommended to set IS size as the block size. Moreover, if we specify the IS size to be false, then the whole file will be considered as one IS and processed by one map task. This takes more time to process as data blocks are residing in various DNs and required to move to the node where map task is running. There are use cases that require to process the whole input file by one map task. Example: XML parser needs to parse entire XML file at the same time.

What if there is a job to process 1000s of files whose size is smaller than the block size?

Hadoop works better with a small number of large files than a large number of small files. A file over 64 MB is split into multiple blocks. If a file is lesser than HDFS block size, an arbitrary file size itself is the block size. There should be at least one block for any size of the file. Therefore, if there are 1000 small files (lesser than 64 MB), then the number of blocks is 1000. Each block will derive one IS by default, which results in 1000 map tasks. This requires huge resources for launching 1000 map tasks. Moreover, scheduling and spawning a map task take more time than processing a small block. Therefore, job latency is high. It is because each map task goes through a sequence of steps in the MR execution sequence. This is the reason why Hadoop works better for a smaller number of large files than a large number of small files. To solve this, we can pack many smaller files (lesser than block size) into one IS using CombineFileInputFormat. Another option is to change IS size.

key-value pair

In MR programming environment, no data can stand on its own. Every data is associated with a key. A key-value pair is called a record. IS and records are logical entities used at the time of job execution. They do not affect physical blocks. Every step in the MR execution sequence takes a key-value pair as input and outputs key-value pairs, as shown in Figure 2.18.

File Input Format

The file input format is responsible for forming IS and preparing records from the contents of blocks. Every file input format defines what should be

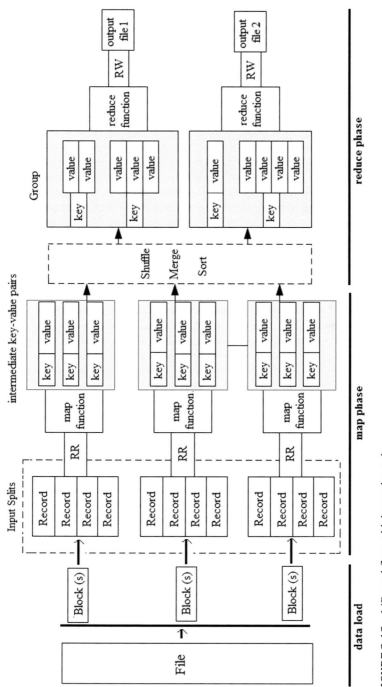

FIGURE 2.18   MR workflow with key-value pair.

the datatype for key and value. Some of the file input/output format and their subclasses are given below. Table 2.1 lists out key-value pairs of different file input format.

InputFormat
1. FileInputFormat
   a. CombineFileInputFormat, CombineSequenceFileInputFormat, CombineTextInputFormat
   b. TextInputFormat
   c. KeyValueTextInputFormat
   d. NLineInputFormat
   e. SequenceFileInputFormat, SequenceFileAsBinaryInputFormat, SequenceFileAsTextInputFormat, SequenceFileInputFiler

2. ComposableInputFormat
   a. CompositeInputFormat

3. DBInputformat

OutputFormat
1. FileOutputFormat
   a. TextOutputFormat
   b. SequenceFileOutputFormat, SequenceFileAsBinaryOutputFormat

2. NullOutputFormat
3. DBOutputFormat
4. FilterOutputFormat
   a. LazyOutputFormat

TABLE 2.1 FileInputFormat and Its Key-Value Type

File Input Format	Key	Default value
TextInputFormat	byte offset	all texts until new line (/n)
KeyValueTextInputFormat	text until first tab (customizable)	all texts after the first tab until new line (/n)
NlineInputFormat	byte offset	texts of "N" lines
TableInputFormat (HBase)	row key	entire row
WholeFileInputFormat	Null	entire file
MultiFileInputFormat	per path basis	per path basis
SequenceFileInputFormat	User-defined	User-defined

Consider an IS comprising two blocks:

Block 1: aa bb cc dd ee ff gg hh ii jj
Block 2: ww ee yy uu oo ii pp kk ll

Map task reads block1 data from aa to jj and does not know how and from where to read block2. That is, map task does not know how to process different blocks as they are physically separate. It is difficult to set an offset that spans across multiple blocks manually. Here comes IS to group block1 and block2 as a single logical block by forming an offset based on the file input format. IS prepares key:value pairs (records) from the content of blocks based on the file input format, as shown in Figure 2.18. What happens when a record spans the block boundary, as shown in Figure 2.19? 5th record from block1 and 10th record from block2 cross its block boundary to the next block.

FIGURE 2.19   Record that spans over block.

HDFS does not know which record is inside the block or spans to the next block. IS figures out where the first record in a block begins and where the last record in a block ends. If the last record in a block is incomplete, IS includes the location information of the next block and the byte offset of the data needed to complete the record.

IS in short,

- Block is a physical entity and HDFS concept.
- IS and record are logical entities and MR concept.
- IS logically groups a number of physical data blocks processed by one map task.
- IS contains information such as the location and size of blocks, not the data of physical blocks.
- IS prepares the content of blocks as key:value pairs based on the file input format.
- You can customize the block and IS size to decide the number of map tasks.
- More than one block in an IS decreases data locality and increases latency.
- IS sets an offset for records that span across blocks.

Step 2: Record Reader (RR)

Like other programming language, MR programming also has data types to deal with different types of data. While InputFormat and OutputFormat deal at the file-level, key-value deal with data-level. Both key and value are declared with a data type. Table 2.2 accounts the data types used in Java and MR.

String in java uses UTF8 or UTF16 bit encoding format, and it is immutable. Text in Hadoop uses UTF8 encoding format, and it is immutable except NullWritable. IS prepares key:value pairs (records) from the content of a block based on file input format and gives a byte-oriented view. The map task does not know how to read those records from IS. Therefore, RR is used to read <key, value> pairs from IS and converts byte-oriented view of records to Hadoop data types and then feeds to the map function. Therefore, the data type of key and value in RR is determined based on a file input format. RR reads a record that crosses a block boundary with offset information from IS. So, the programmer need not worry about it. Each InputFormat provides its RR implementation to read key/value pairs. Example: for TextInputFormat, IS forms byte offset as key and entire line (until new line) as value. RR will convert the key (byte offset) to LongWritable and value (entire line) to Text, as shown in Table 2.3, and feed to map function.

TABLE 2.2   Data Types in Hadoop MR

Java primitive type	MR data type
Boolean	BooleanWritable
Byte	ByteWritable
Byte[]	BytesWritable
Short	ShortWritable
Integer	IntWritable VIntWritable
Float	FloatWritable
Double	DourbleWritable
Java objective type	**MR data type**
String	Text
Object	ObjectWritable
	NullWritable

TABLE 2.2 *(Continued)*

Java primitive type	MR data type
Java collection	**MR data type**
Array	ArrayWritable
	ArrayPrimitiveWritable
	TwoDArrayWritable
Map	MapWritable
SortedMap	SortedMapWritable
Enum	EnumSetWritable

If you create a new file input format, it is your responsibility to define IS, RR, and check for record boundary. So, you have to define the data type for input key and input value for map function accordingly. MR data types are highly optimized than java data types to send/receive data over the network. If you do not use MR data type, you need to serialize and de-serialize manually whenever data is distributed. Table 2.4 shows the size of serialized data types.

It is not possible to perform arithmetic operations (+–/*%) with MR data type. So, if you want to do it, you have to convert them to java primitive type. Before Java1.4, conversion between objective and primitive types were done using parse***(), toString(), valueOf() functions. In later versions, the conversion is done automatically using autoboxing and unboxing. However, there is no autoboxing and unboxing to convert data type between java primitive/objective and MR data types. Therefore, we have to convert manually between java primitive/objective and MR data types to perform any meaningful arithmetic operations. ***Writable() constructor wraps the primitive/objective type of java into MR data types.

TABLE 2.3 Data Types for FileInputFormat

Wrapper classes (in java)	Primitive types (in java)	Box classes (in MR)	Serialized size (in bytes)
Boolean	Boolean	BooleanWritable	1
Byte	Byte	ByteWritable	1
Short	Short	ShortWritable	2
Integer	Int	IntWritable	4
		VIntWritable	1–5
Float	Float	FloatWritable	4
Long	Long	LongWritable	8
		VLongWritable	1–9
Character	Char	Text	2 GB
String	Character array		

TABLE 2.4   Size of Serialized Data Types

File Input Format	Key	Data type	Default value	Data type
TextInputFormat	Byte offset	LongWritable	Until new line	Text
KeyValueTextInput-Format	Data until first tab	Text	Until new line after first tab	Text
NLineInputFormat	Byte offset	LongWritable	N lines	Text
WholeFileInputFormat	Null	NullWritable	File contents	Text

Syntax: ***Writable temp=new ***Writable (*** java primitive type)
Example: int a=10;
        IntWritable num=new IntWritabl (a);
        String str="I am learning hadoop";
        Text t=new Text(str);

To perform any numeric/string manipulation, Hadoop data types are not suitable (lack of utility functions). Therefore, we have to convert Hadoop box class data type to java primitive/objective type and then perform manipulations inside map/reduce functions. To convert from Hadoop box class data type to java primitive type:

IntWritable num=new IntWritable (10);
int a=num.get();
Text t=new Text("I am learning hadoop");
String str=t.toString()

get() method is used to convert integer MR data type to java primitive data type. toString() is used to convert character/string type of MR to java string type.

Step 3: Map task

Mapper or map task is an user-defined map function, which is mostly used for pre-processing records. Therefore, the output of the map task is called as intermediate key-value pairs, but not the result of MR Job. RR feeds a record to map function. Setting data type for map input key and map input value is based on the output record of RR. But, the data type for map output key and map output value can be customized. When a function takes input and produces output, it is called as task. As in Figure 2.20, when a map or reduce function takes input and produces output, they are called tasks: map task and reduce task. In general, map task is called as mapper, and reduce task is called as reducer. Map and reduce tasks are invoked only once at the time of launching. But, map and reduce function are invoked for every input record.

Every map task has its RR, as shown in Figure 2.15. If there are "n" input records in IS, "n" times map function is invoked and produces zero or more output records. The map function lets users write their logic to decide what to do with the records. You can parse the value and extract only relevant fields (projection) or filter unwanted/bad records or transform the incoming records. Output of map task is stored in an in-memory buffer, as shown in Figure 2.23. From Figure 5.12, you can observe that the map phase ends only after all the map tasks of a job are completed. Each map task latency may vary due to other simultaneous activities in the system.

Step 4: Partitioner (balancing reduce task input)

If more than one reduce task is launched, partitioner decides to which reduce task a map output record should go. The partitioner is a function that balances reduce task input size from all map tasks. The partitioner is meaningful only when we launch more than one reduce task. Partitioner splits the output of a map task into several partitions. A partition is a portion of map output that goes to a particular reducer. The number of partitions is equal to the number of reduce tasks. The objective of partitioning is to bring the same key from different map tasks into one reducer. For instance, consider 10 map tasks for a wordcount job. Consider map task 1, 4, 6, and 10 produce the word "Hadoop" as output. To find the total count of the word "Hadoop," it should be brought to one reduce task. Therefore, the word "Hadoop" can be counted as 4.

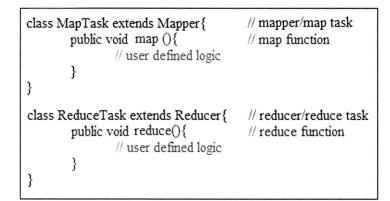

```
class MapTask extends Mapper{          // mapper/map task
        public void  map (){           // map function
                // user defined logic
        }
}
class ReduceTask extends Reducer{      // reducer/reduce task
        public void reduce(){          // reduce function
                // user defined logic
        }
}
```

FIGURE 2.20 Function vs task.

Default partitioner is HashPartitioner, which calculates the hash value to decide to which reducer current record should be sent. It works well with any number of partitions and ensures each partition has a right mix of keys, leading to more evenly sized partitions. Sometimes, it might cause skewness. We can also customize partitioner to decide which record should go to which reducer. There are different in-built partitioners:

BinaryPartitioner – partitions binary data
KeyFieldBasedPartitioner – taking any field of a record to the partition
TotalOrderPartitioner – helps to equally distribute the map outputs to the reducers

Step 5: Combiner (network and disk IO optimization)

Combiner task minimizes network traffic, disk IO transfer, and the number of records processed by reduce task. If the map task generates huge output, it leads to more traffic on the local network to transfer them to reduce task. Secondly, until map output is transferred over the network, it is stored in an in-memory buffer (refer Section 2.9 to understand in-memory buffer). When map output size goes beyond the in-memory buffer threshold, it should be spilled into the local file system in the disks. It requires more disk IO transfer. Finally, a reduce task will have to process huge data so that the latency will be high. In order to solve these issues, the combiner function is used to minimize the size of map output locally after partitions after prepared. This gives very less size of data to spill into local disk and transfer over the network to reduce task. So, the combiner is called as mini/local reducer. If there is no way to minimize the map output, there is no use of enabling combiner. Therefore, the combiner is purely application specific. When map output size goes beyond the in-memory buffer threshold, the combiner is applied onto map output and spilled into the local file system. Each map task has its combiner to execute.

How many times is the combiner task executed?

After map output is partitioned, the combiner task is launched once for each partition sequentially as a separate thread along with map task. If there are four partitions in current map output, four times combiner task is launched, and combiner output is spilled into local disk. After the map task is over, there can be several spill files in local disk over some time. If there are at least three spill files (min.num.spills.for.combine – default is 3), respective partitions from each spill is merged and sorted. Combiner task is launched once for each merged partition. If there are just two spills from a map task, the potential reduction in map output size is not worth than the overhead of invoking combiner. Therefore, the number of times the combiner task launched is not fixed. If the combiner is enabled, it is launched at least once in each map task. However, the number of times combiner task launched (if enabled) is

= (number of partitions *number of spills) + number of partitions.

Combiner function can be defined by the user or can be the same as reduce task. Signature of the user-defined combiner task is identical to the reduce task. However, all applications cannot use reduce task as combiners. Application that satisfies additive and commutative property can use reduce task as combiner task. Example: sum, max, min.

Commutative: a + b = b + a (swapping operands result the same)
Associative: a + (b + c) = (a + b) + c (grouping different operands result the same)

Example:
1. Max function satisfies: max (max(a,b), max(c,d,e)) == max(a,b,c,d,e)
2. Mean function does not satisfy: mean(mean(a,b), mean(c,d,e))!= mean(a,b,c,d,e)

For mean function, associative property does not hold. So, the reduce function cannot be used as a combiner function for a MR job to find mean. Combiner is mainly used for jobs like sum, count. It is not automatically invoked, and it is our responsibility to invoke combiner to execute.

2.6.2   REDUCE PHASE

Reduce phase also comprises a sequence of steps to be carried out after the map phase ended. The steps are: shuffle → merge → sort → group → reducer → file output format → record writer. This phase includes the magical steps (shuffle + merge + sort + group) that show the power of Hadoop MR. Magical steps are executed only if reduce task is executed. These are completely managed by the MR framework itself. However, programmers can customize to suit their application. Please note that shuffle, merge, sort and group are logical operations, and they do not change the original output of mappers. Reduce phase begins when some percentage (mapred.reduce.slowstart.completed.maps) of map tasks are completed. But, it does not mean that reduce function is executed. Reduce function in each reduce task is executed only after all the map tasks are completed. In general, reduce phase moves output (partitions) of mapper/

combiner from map nodes to reduce nodes, merges all the partitions, sorts based on the key, and groups all the values that belong to the same key to eliminate key redundancy. Finally, the output (key:list(values)) of group function is fed to reduce function.

Step 6: Shuffle (copy)

Shuffle is the process by which partitioned mapper/combiner output is transferred over HTTP network to one or more TT where reduce tasks will be executed. This is also called as copy phase. Each reduce node receives one or more partitions from all map tasks. If reduce task is decided to run in the same node where map task is completed, shuffle does not have any role. Each reduce node uses five threads (mapred.reduce.parallel.copies) by default to pull its partitions in parallel from the map nodes. But, how does a reduce node know which map node to query for partitions? This is done with the help of JT. As each map task completes, it notifies the JT about partitions. Each reducer periodically queries JT to know the node running map tasks. Consider 2 TB input file. What if the mapper output size is the same as its input? Moving 2 TB to reduce nodes across the network requires huge bandwidth. That is the reason the combiner is run to minimize the size of map output to move across a network to reach reduce node. To minimize the size of map output further, it can be compressed. Snappy is the most commonly used compressor in MR. Other compression schemes are DEFLATE, LZO, LZ4, etc. Refer Section 6.12 for more information on compression.

Step 7: Merge and Sort

Respective map output partitions are copied to the reduce task Java Virtual Machine (JVM). If memory is not sufficient, they are spilled into local disk. As soon as partitions from all map tasks arrived to reduce nodes, partitions should be merged into one file for further processing. Every 10 (io.sort. factor) spilt files are merged by default. The merged file should be sorted based on key.

Step 8: Group

A set of values that belong to the same key is grouped to eliminate key redundancy. The number of times reduce function executed is equal to the number of key:list(values) pairs after grouping. If duplicate keys exist, reduce function is repeatedly invoked for every duplicate key. If the same key is present in multiple spilled files, for a single key, many files have to be brought into memory (needs multiple passes) to feed reduce function. This involves a lot of IO. That is the reason, we sort and group keys before calling reduce function. Therefore, only one pass is required for each key, and only one time, reduce function is invoked for each unique key.

Step 9: Reduce task

Reduce function processes a list of values for each key and produces zero or more output records. The number of times reduce function executed is equal to the number of records (key: list of values) after grouping. The map function is mainly used for pre-processing the records. However, the core algorithm is implemented in reduce function. Aggregation and join operations are performed here. Map output is deleted after successful completion of all reducers. If a node running map task fails before the map output is taken by reduce tasks, the framework will automatically rerun the map task to create map output again.

JT decides in which node reduce tasks should be executed. It could be the same node where map tasks executed or some other node in the same rack or node in some other rack. It depends on the slot availability and the load of the local network. Users can specify the number of reduce tasks based on the requirement of parallelism. If not specified, a default reduce task is launched, which consumes map output and results as it is, but in a sorted manner as it passes through the shuffle, sort, and group functions. If you set the number of reduce tasks is zero, map output itself is considered to be job output and will be stored in HDFS. There will be no shuffle, sort, merge, and group process for this case.

As you increase the number of map and reduce tasks, data can be processed in more parallel, so job latency is minimized. However, it requires more resources. There should not be too many reduce tasks doing very little work and a few doing more work. Reduce tasks do not have the advantage of data locality, unlike map tasks, because it has to receive partitions from various mappers running in different nodes. However, it is rack-aware. The output of reduce task is generally stored in HDFS by default with replication unlike map output stored in memory or spilled in the local file system. The first copy is stored locally where reduce task is running, the second copy is stored in any node in the same rack, and the third copy is stored in any node in some other rack in the cluster. Therefore, writing reduce output consumes network bandwidth as much as normal HDFS write pipeline consumes.

Key points

- Parsing, filtering, and record-level transformations can be done in the map function. Do not perform any aggregation kind of operations in map task. Because the objective of the map task is not aggregating.
- Operations like grouping, sorting, aggregation, and join should be implemented in reduce task. Do not do parsing, filtering, etc. in reduce function even though it is possible. Because the objective of reduce task is not to perform filtering.
- If your application does not require sorting, grouping, and aggregation, reduce task is not going to be useful. Some applications will require only map tasks for pre-processing.
- The number of mappers is determined by the MR framework, and the number of reduce tasks is decided by users.
- If no reduce task is specified, a default reduce task is run, which outputs what it consumes.
- Input key, input value datatype for reduce function should be the same as output key, output value datatype of the map function.
- Output key and output value data type of reduce function can be defined by the user.
- Input and output records of map function are not in sorted order. However, the input records of combiner and reduce function is always sorted order.

By default,

- number of blocks == number of IS == number of map tasks
- number of partition == number of reduce tasks == number of output file.

Step 10: Record Writer (RW)

The reduce function feeds output key and output value to RW, which in turn writes onto HDFS with tab separation by default based on TextOutputFormat. Each reducer writes an output file onto HDFS with the desired RF. RW opens an output file and writes the reduce output records. The output of map tasks is spilled into local file system because writing intermediate results onto HDFS will lead to unnecessary replication and leads extra work to delete them later. However, the output of reduce task is written onto HDFS as it is the final result and should be fault-tolerant. The naming convention of output files is part-r-00000. Part denotes the part of an output file, "r" indicates the output of reduce task, and five digits indicate output file number. There can be 1 lac reduce output files created as a result of a MapReduce job.

2.6.3   DAEMONS THAT EXECUTE MR STEPS

Figure 2.21 shows the daemons that control the steps in MR execution sequence. NN handles when big data is uploaded and split into equal sized blocks. JT handles job life cycle management and forming IS. From RR to combiner function is executed in the map task JVM. The magical steps (shuffle, merge, sort, group) are handled by MR execution framework itself. The remaining steps are executed by reduce task to finish job execution. Users can define almost all functions in the execution sequence, but only map, combiner, partitioner, and reduce functions are quite commonly customized. There is a driver class in MR job, from which all these user-defined functions can be invoked.

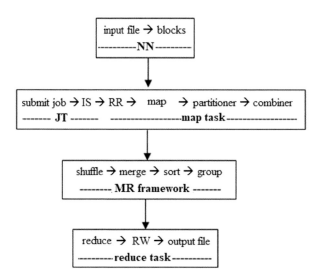

FIGURE 2.21   Daemons managing MR steps.

2.7   MR EXAMPLE – I

Based on the comprehensive review of MR, Figure 2.22 illustrates the word-count example. Our objective is to get the total frequency of each word from the given input file. Consider an input file with three blocks. Each IS points to one block. Map function tokenizes the input records and assigns value "1" for each word. With the help partitioner, words that start with A to C are sent to reducer 0, and the rest of the words are sent to reducer 1 in shuffle phase. The collected words from different files is merged into one file. Then, records are sorted and grouped based on key. Therefore, for each key, value 1 is clubbed. Finally, reduce function sums up the list of "1" assigned to individual word. Each reduce task will generate an output file onto HDFS with replication.

2.8   MR EXAMPLE – II

2 GB of sales information is given. We need to find the total money spent by each customer. We assume the following input file and one reduce task for the job.

Name	Price	Item	Date
Sham	7000	mobile	12-2-2014 ....
Ravi	1000	earphone	10-2-2015 ....
Raja	10000	laptop	2-1-2015 .....
Anil	2000	charger	3-2-2014 ....
Bala	1000	books	5-3-2011 ....
Ravi	500	pizza	23-11-2015 ....
Sham	100	soap	12-12-2015 ....
Anil	2000	charger	3-2-2014 ....
Ravi	1000	snacks	5-3-2011 ....

As already discussed, there are two phases in the MR execution sequence. Each phase consists of a sequence of steps taking input and output in the form of key-value pairs.

Map phase

Step 1: file → blocks → FileInputFormat → IS → RR

Step 2: RR (k1, v1) → map

Step 3: map (k1, v1) → (k2, v2) partitioner

Step 4: partitioner (k2) → gives reduce task number → spilling process → combiner

Step 5: combiner (k2, list(v2)) → (k3, v3) → shuffle

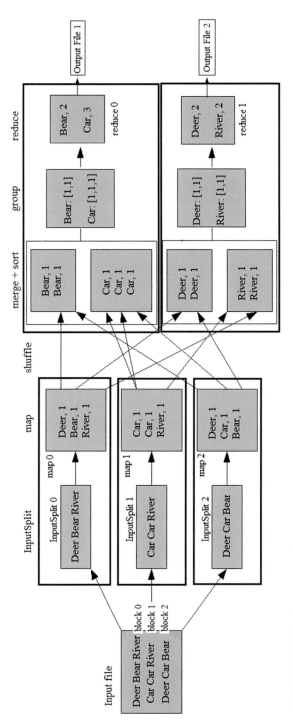

FIGURE 2.22 Wordcount example.

Reduce phase

    Step 6: shuffle → merge → sort
    Step 7: sort (k3) → (k3, v3) → group
    Step 8: group (k3) → (k3, list (v3)) → reduce
    Step 9: reduce (k3, list (v3)) → (k4, v4) → OutputFileFormat → RW
    Step 10: RW (k4, v4) → output file

Note: while moving from one step to another step, if key:value is modified, then (k#, v#) is incremented by 1. Example: key and value after map are incremented from (k1, v1) to (k2, v2).

2.8.1   MAP PHASE

Step 1: file → blocks FileInputFormat → IS → RR

The user submits 2 GB sales file to HDFS client, which in turn divides the file into 32 blocks (2GB/64MB). Then, HDFS client communicates NN to upload these data blocks onto HDFS. NN assigns a blockID and gives a list of three DNs for each block. For better understanding, the block is not replicated in this example and consider only two blocks (B1, B2) are stored in node1 and node2 respectively for further discussion.

B1 in node1

Sham	7000	mobile	12-2-2014 ….
Ravi	1000	earphone	10-2-2015 ….
Raja	10000	laptop	2-1-2015 …..
Anil	2000	charger	3-2-2014 ….

B2 in node2

Bala	1000	books	5-3-2011 ….
Ravi	500	pizza	23-11-2015 ….
Sham	100	soap	12-12-2015 ….
Anil	2000	charger	3-2-2014 ….
Ravi	1000	snacks	5-3-2011 ….

Once DNs received data blocks, BR is sent to NN to update meta-data. Then, NN sends an success acknowledgment to HDFS client. Now, write MR job to find total money spent by each customer and submit to MR job client, which communicates to JT to get a job ID, determines IS (2 here), and copies all job-related information onto HDFS. Job client now submits the job to JT, which prepares an execution plan. Map task is launched in the nodes where the required data block is available.

Step 2: RR (k1, v1) → map

RR is used to read the records (key:value) from IS. By default, TextInput-Format is used to read the records. Therefore, the key is byte offset, and value is the entire line (until newline). Preparing records in IS does not mean that key is attached to each line. Keys are created only when RR reads a line. Following is the output of RR reading from IS.

node1:	byte offset	name	price	item	date
	0	Sham	7000	mobile	12-2-2014 ....
	70	Ravi	1000	headset	10-2-2015 ....
	100	Raja	10000	laptop	2-1-2015 .....
	150	Anil	2000	charger	3-2-2014 ....
node2:	byte offset	name	price	item	date
	2000	Bala	1000	books	5-3-2011....
	2100	Ravi	500	pizza	23-11-2015 ....
	2250	Sham	100	Soap	12-12-2015 ....
	2431	Anil	2000	charger	3-2-2014 ....
	2599	Ravi	1000	snacks	5-3-2011 ....

Byte offset is a count of the number of characters (each 1 byte size) in the previous line + 1. Example: the first-row key starts with 0 characters. The second-row key is the number of characters in the first row + 1. The third-row key is the number of characters in the first two rows + 1. We have given random byte offset as we are reluctant to calculate the number of characters in each line manually.

You can observe that the byte offset across IS is unique using TextInput-Format. Byte offset does not start at 0 in every IS. Since offset is byte-based, it is easy to calculate the starting byte offset of the second IS just by knowing the size of the first IS. Therefore, it is unique across the IS.

You might wonder why we cannot use line number as an offset. We cannot calculate the number of lines in IS by just having the size of IS. Moreover, the second IS has to wait until the first IS counts the number of lines. So, second IS key can be incremented from there on. Using byte offset, it is straightforward to maintain a global offset for key across IS. If the file input format is KeyValueTextInputFormat, the key is any data that appears till before first tab character and value is the entire line (until newline) after first tab character.

Step 3: map (k1, v1) → (k2, v2) → partitioner

RR gives the details of every customer transaction line by line to the map function. Particularly, our program filters out only the customer name and money paid for each transaction. Therefore, map function provides output (*key*-customer name, *value*-price). As we are going to calculate the total money spent by each customer, we have chosen customer name as the map output key. Map tasks in node1 and node2 buffer their output in local memory. If memory is filled up, it is spilled into the local file system with file name part-m-00000 in node1 and part-m-00001 in node2. "m" indicates output is from map task. The output of map function in both nodes:

node1		node2	
part-m-00000		*part-m-00001*	
Sham	7000	Bala	1000
Ravi	1000	Ravi	500
Raja	10000	Sham	100
Anil	2000	Anil	2000
		Ravi	1000

Step 4: partitioner (k2) → gives reduce task number → spilling process combiner

Only one partition is formed from map output as we are going to launch only one reduce task. The partitioner gives reducer number as output if there are more than reduce tasks. There will be no change in key-value pairs in the partition process.

Step 5: combiner (k2, list(v2)) → (k3, v3) → shuffle

Combiner is explained at the end of this chapter. Further explanation is done with the output of map task, not the output of combiner for more clarity. However, you can follow the same procedure with combiner output also.

2.8.2 REDUCE PHASE

Step 6: shuffle → merge → sort

Shuffle moves the partitions from map tasks to the node running reduce task (say node2) over the network. JT decides reduce node based on the network traffic and slot availability. So, partition from map task1 in node1 is moved to node2. Now, both partitions are merged and the output of merge process in node2 is given below.

node2

Sham	7000
Ravi	1000
Raja	10000
Anil	2000
Bala	1000
Ravi	500
Sham	100
Anil	2000
Ravi	1000

Step 7: sort (k3) → (k3, v3) → group

After merging, the data is sorted based on key (customer name) and maintained in node2 itself. The output of sorting is as below.

node 2

Anil	2000
Anil	2000
Bala	1000
Raja	10000
Ravi	1000
Ravi	1000
Ravi	500
Sham	7000
Sham	100

Step 8: group (k3) → (k3, list (v3)) reduce

After sorting, it is easy to group all the values that belong to the same key (customer name). The output is,

node 2

Anil	[2000, 2000]
Bala	[1000]
Raja	[10000]
Ravi	[1000, 1000, 500]
Sham	[7000, 100]

Step 9: reduce (k3, list (v3)) → (k4, v4) → OutputFileFormat → RW

As we aim to find the total money spent by each customer, we are going to add all the values that belong to each key. The output of the reduce function is,

node 2

Anil	[4000]
Bala	[1000]
Raja	[10000]
Ravi	[2500]
Sham	[7100]

Step 10: RW (k4, v4) → output file

The final output is written onto HDFS with file name part-r-00000. Output key and value are written with tab space as delimiter based on TextOutputFormat. The client is notified about the job completion, and job statistics are displayed on the console.

node 2

part-r-00000

Anil	4000
Bala	1000
Raja	10000
Ravi	2500
Sham	7100

The delimiter "tab space" can be changed to any other symbol. Example: to change the delimiter to comma, you have to use mapred.textoutputformat. separator property.

2.8.3 DISK AND NETWORK IO OPTIMIZATION

Combiner function is run right after partitioning process and before spilling map output into a disk. The objective is to minimize the output size before spilling into a disk or moving onto a network to reach reduce node.

Step 5: combiner (k2, list(v2)) → (k3, v3) → shuffle

The following steps for combiners are very similar to original MR execution flow. Refer to Section 2.9 to know what is happening while applying combiner function. Output of each step is given below:

map (k1, v1) → (k2, v2) → partitioner
partitioner (k2) → local sort
local sort (k2) → (k2, v2) → local group
local group (k2) → (k2, list(v2)) → combiner
combiner (k2, list(v2)) → (k3, v3) shuffle

map (k1, v1) → (k2, v2) → partitioner

node1		node2	
Sham	7000	Bala	1000
Ravi	1000	Ravi	500
Raja	10000	Sham	100
Anil	2000	Anil	2000
		Ravi	1000

partitioner (k2) → local sort

node1		node2	
Anil	2000	Anil	2000
Raja	10000	Bala	1000
Ravi	1000	Ravi	500
Sham	7000	Ravi	1000
		Sham	100

local sort (k2) → (k2, v2) → local group

node1		node2	
Anil	[2000]	Anil	[2000]
Raja	[10000]	Bala	[1000]
Ravi	[1000]	Ravi	[500, 1000]
Sham	[7000]	Sham	[100]

local group (k2) → (k2, list(v2)) → combiner

node1		node2	
part-m-00000		*part-m-00001*	
Anil	[2000]	Anil	[2000]
Raja	[10000]	Bala	[1000]
Ravi	[1000]	Ravi	[1500]
Sham	[7000]	Sham	[100]

combiner (k2, list(v2)) → (k3, v3) → shuffle

Go to step 5

2.9   ANATOMY OF MAP AND SHUFFLE

Map output (output_key, output_value) is written into circular in-memory buffer (defined by io.sort.mb) as shown in Figure 2.23. This buffer size is 100 MB by default in the map task JVM. Whenever this circular buffer is almost full (io.sort.spill.percent – by default 80%), a spilling thread starts in parallel to the running map task. Every map task has its combiner and partitioner in the map task JVM itself. Note that if the spilling thread is too slow and the buffer is 100% full, map function cannot proceed and it has to wait. There is a sequence of steps carried out between map task and shuffle: Map output → partition → local sort → local group → combiner → spill → shuffle.

1.   Map task writes output records (key-value) into a circular in-memory buffer.

2.   Once buffer is filled up 80%, a background thread starts in parallel to map task and partitions map output into several segments. Number of partitions is equal to the number of reduce tasks. Partitioner assigns partition ID for each map output record. User can customize the partitioner by setting the configuration parameter mapreduce. partitioner.class.

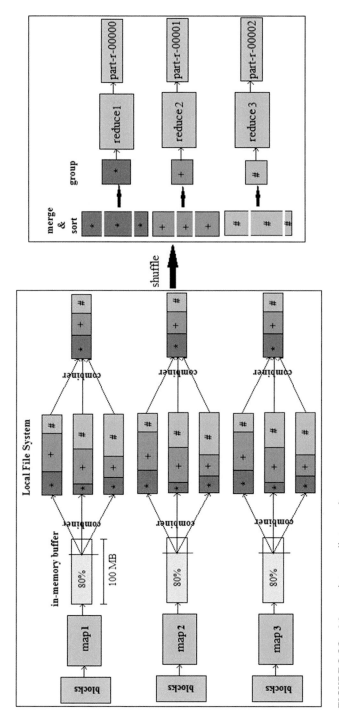

FIGURE 2.23 Mapper intermediate results.

3. Each partition is sorted in memory based on triplets (partitionIdx, map_output_key, value).

4. Values are grouped based on the map output key to eliminate key redundancy in each partition.

5. If the combiner is enabled, it is launched once for each partition. Combiner function receives (key:list(values)).

6. All partitions are spilled into the local file system (location specified by mapred.local.dir property).

7. At the end of map execution, there can be many spilled files at mapred.local.dir. If there are less than 3 spills, partitions are sent directly to reduce nodes. If there are 3 or more spills (min.num.spills. for.combine – default is 3) in the disk, then spill files are merged (io. sort.factor – by default 10 files at the time) into one file.

8. A single merged file contains so many partitions, which are sorted based on (partitionIdx, map_output_key) once again.

9. All values that belong to a key (partitionIdx, map_output_key) in the partition are grouped to remove key redundancy.

10. Combiner task is launched once again for each partition.

11. Each partition is copied over HTTP network to reduce nodes.

If map produces no output or less than three spills, combiner task execution is not guaranteed even if it is enabled because the invocation of combiner adds time overhead than processing map output records. Therefore, combiner task is executed as a thread for 0 or 1 or any number of times without affecting the final result.

Why map output is written into the local file system, but not onto HDFS?

Writing onto HDFS is based on blocks and involves replication. So, if a map task is killed or failed, there will be lots of intermediate files sitting on HDFS for no reason, which you will have to delete manually. If this happens too many times, the cluster performance will degrade. HDFS is optimized for appending and not frequent deleting. Therefore, so much operation is involved just for a failed job cleanup process for no gain. Optionally, map output can be compressed to minimize the size of spill files and network IO.

Job metrics

Upon job completion, JT displays the statistics of job on the console. These statistics are collected using counters. There are two types of counters: built-in counter (job, task, and file-level), and user-defined counter. TT and DN aggregate all counters and send it to their masters. This information is retained in job history for later use. Counters are used to determine data quality, diagnose, and debug problems. It is implemented using Enum. Users can create counters to enumerate something of interest. Counters can be accessed via WUI, Java API, and CLI. Several built-in counters are:

- Job counters: number of map and reduce tasks launched, number of failed tasks, etc.
- File system counters: number of bytes read and written, etc.
- MR framework counters: network, and memory-related statistics, etc.

Apart from job metrics, task and job completion status is also displayed in the output console. This is discussed in Section 2.12.

2.10   MR-RELATED TERMINOLOGY REVIEW

NN: Manages file system meta-data, controls and coordinates DNs.

DN: Manages local disks, stores, and retrieves data blocks upon NN instructions.

SNN: Maintains a copy of meta-data from NN to avoid SPOF.

JT: Prepares the execution plan, launches map/reduce tasks, and performs phase coordination.

TT: Manages CPU and memory as slots, runs map/reduce tasks, tracks task status, and reports to JT.

HDFS Client: A daemon that receives data upload request and interacts with NN to place data blocks. It can be run in any nodes in the Hadoop cluster.

Job Client: A daemon that receives MR job and interacts with JT. It can be run in any nodes in the Hadoop cluster.

Job: It is a logical pack of the following things: input data, MR program, and configuration information. A job can have any number of map and reduce tasks.

Task: A piece of code running within a single thread of execution is called a task in general.

Map task: It executes a user-defined map function, which reads data from HDFS as key/value pairs and produces an arbitrary number of intermediate key-value pairs.

Reduce task: It executes a user-defined reduce function, which collects the output from all map tasks, merges, sorts, groups values that belong to the same key, and finally produces an arbitrary number of key-value pairs.

Driver program: A program that initiates and controls the execution of map and reduce tasks.

2.11 MR EXECUTION FLOW

So far, we have been getting more insight on MR job execution sequence. Now, let us understand how MR jobs are carried out. Major steps are given in Figure 2.24. Detailed explanation is given in Figure 2.25.

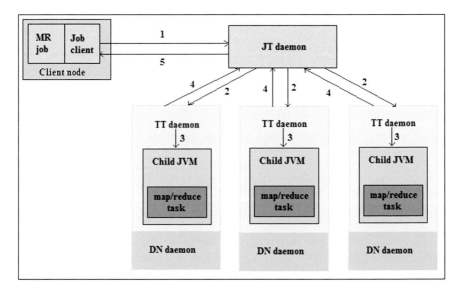

FIGURE 2.24 MR job execution flow.

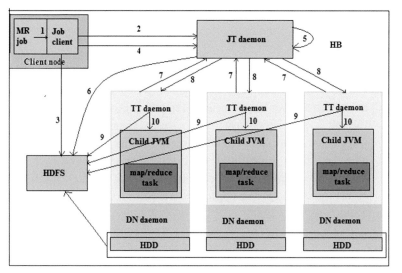

FIGURE 2.25 Detailed MR job execution flow.

1. A user submits a job to Job client, which in turn submits the job to JT.
2. JT prepares the execution plan, distributes tasks, and coordinates MR phases.
3. TT launches map/reduce tasks.
4. TT sends progress of map/reduce tasks and HB to JT.
5. JT sends job progress status and completion message to the client.

Job submission

1. Upload data onto HDFS before you launch an MR job. Then, the user submits a job to Job client.
2. Job client goes for various checks, whether the input file and output directory in HDFS exist by interacting NN. If the input file does not exist and/or output directory is already available, the exception is thrown, and the job is terminated. Otherwise, Job client interacts with JT to get jobID and calculates IS.
3. All job-related information such as job.xml, job file, and IS information are copied to 10 DNs by default to be highly available during execution.
4. Once all information is moved onto HDFS, Job client submits a job to JT to begin the job life cycle, and the output directory is created.

However, the output files from reduce tasks are created only at the end of the job life cycle.

Job initialization

5. Calculating IS by Job client may be slower. So, you can set JT to determine IS by communicating NN. Once this is done, the job is placed in the job queue. Then, MR job scheduler picks up the job from the job queue and creates an object to encapsulate its map/reduce tasks.
6. JT retrieves IS information from HDFS (if calculated by Job client) and prepares an execution plan based on the locality of data blocks. Every block has three copies by default. So, JT prepares a list (for block1, say, DN2, DN1, DN3) for each block. DN2 comes first in the list as it is very near to JT due to rack awareness.

Task assignment

7. TT runs an infinite loop that periodically sends HB and task status information to JT.
8. The scheduler attempts to launch a map task on block1 in TT2. If there is no free slot in TT2, TT1 is attempted. If there are no free slots in the three TTs mentioned for block1 to achieve data locality, block1 is copied to any other TT which has a free map/reduce slots to perform non-local execution.

Task execution

9. TT localizes job-related information such as jar file, and job.xml file from HDFS.
10. TT launches a JVM for running map/reduce tasks. NN instructs DN to bring the required block from HDFS and loads into JVM.

Once all map and reduce tasks are done, JT updates the job status as success and notifies the Job client. In the end, WebUI will present all statistical and runtime information. Finally, JT instructs TT to remove JVM and reclaim resources safely.

2.12   JOB PROGRESS CALCULATION

Job and task progress are periodically displayed in command line interface and web user interface in percentage. Map phase progress is calculated based on the number of map tasks completed per job, as shown in Figure

2.26 (a). Once all map tasks completed (100%), shuffle (copy) starts copying map output from all map tasks. Unlike map phase, reduce tasks individually progress in reduce phase. A reduce task progress is denoted in three different colors, as shown in Figure 2.26.

- Green color signifies shuffle progress.
- Red color signifies merge and sort progress.
- Grey color signifies reduce function progress.

In a reduce task progress,

- 0-33% denotes that shuffle has completed.
- 34-66% means that sort and group have completed.
- 67%-100% indicates that all reduce tasks is completed.
- If only half the number of the reduce tasks completed, then the progress is 1/3 + 1/3 + 1/6 = 83%.

Figure 2.26(b) shows the first reduce task nearing completion (83%) after shuffle, sort, and merge completed. At the time of third and fourth reduce task on copy stage, first and second reduce tasks completed the execution, as shown in Figure 2.26(c). Figure 2.26(d) shows the completion of all reduce tasks (100%).

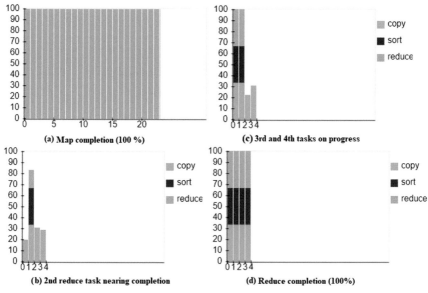

(a) Map completion (100 %)

(c) 3rd and 4th tasks on progress

(b) 2nd reduce task nearing completion

(d) Reduce completion (100%)

FIGURE 2.26   Job and task progress.

2.13   FAILURE HANDLING (FAULT-TOLERANCE)

A system must be highly available and reliable. Servers are more likely to fail if you have many. A server may crash, resulting in data block loss and task failure. Hadoop uses the checksum to detect data corruption and HB to detect node/task failures.

Availability – minimization of system (hardware/software) downtime for a particular period to serve requests.

Reliability – availability + redundancy. Reliability particularly ensures the minimization of data block loss and task failure.

- To mitigate data loss, more than one copy of the same data block is stored in different servers. Therefore, if a server is crashed, a copy of data block from another server is used.
- To mitigate task failure, a copy of the same task is sent to another server which has a desired data block and re-executed.

Speculative execution: A task that is slower than peer tasks and does not make any progress is called straggler. It causes longer latency for a job. To avoid straggler, the same task is executed in some other node while a slower task is still running. The one who finishes early is considered, while the other one is killed. Failures in MR can happen with map/reduce task, TT, and JT, as shown in Figure 2.27. NN blacklists DNs which gives bad records frequently. So, the latest tasks will not be allocated to particular TT.

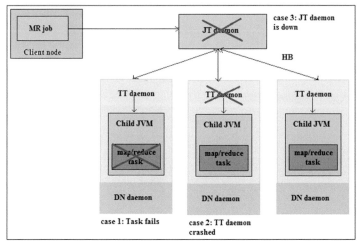

FIGURE 2.27   Failure scenarios.

MR task failure

A map/reduce task failure can happen in different ways.

- Bug in MR function: any task that throws an exception (if not handled) is marked failed.
- Bug in JVM: If the JVM suddenly exits, TT marks the specific task as failed.
- Hanging (due to infinite loop) tasks stop sending progress updates to JT. So, JT kills the JVM and marks that task failed.

Failed tasks are reported to JT by TT. So, JT reschedules the failed tasks to some other nodes except the node where a particular task recently failed. If a task repeatedly failed four times (mapred.map/reduce.max.attempts), it will not be retried again, and the whole job is terminated. If three different tasks of the same job failed, the whole job is considered to be a failure. You can also set when a task should be declared as failed by using mapred.task.timeout (default 0) property.

TT failure

If a slave node or TT daemon crashed or task runs very slowly, JT will not get periodic HB from TT. If there is no HB from TT for 10 minutes, the particular TT will be removed from the cluster. All tasks already scheduled in that TT will be rescheduled to some other TT. Completed map task in the failed TT is re-executed in some other TT. Because, map output has been stored in the local disk of failed TT, so it is not accessible. Completed reduce tasks need not be re-executed as their output is stored in HDFS with replication.

JT manages the state of each job, and the partial result of a failed task is ignored. If different tasks (by default 3) of a job failed in a TT, particular TT is blacklisted by JT for that job, and no other tasks of that particular job are assigned to that TT. However, tasks of other job are executed in that TT. Blacklisting may expire over time, and again, TT may get a chance to run tasks. This can be customized using mapred.max.map.failures.percent property, which is supposed to be less than mapreduce.map/reduce.maxattempts.

JT failure

If JT is down, all jobs running in the cluster is marked as failure. There-fore, JT also suffers from SPOF, and there is no high availability and auto failover. Very similar to NN, there is a possibility to set checkpointing for JT too.

2.14 MR WEAKNESSES AND SOLUTIONS

1. Lack of selective access to input data: For every MR job, entire input data is processed. It is not possible to access a subset of data. There are some tools to achieve this:

    • Indexing data: Hadoop++, HAIL
    • Intentional data placement: CoHadoop
    • Data layout: Llama, Cheetah, RCFile, CI

2. Redundant processing and re-computation: Different jobs perform similar processing on the same input data. There is no way to reuse the results of previous jobs. So, we have to re-compute the same data for future jobs. Some possible tools to overcome this problem are:

    • MRShare
    • Result sharing and materialization: ReStore
    • Incremental processing: Incoop

3. Lack of early termination: In general, all map tasks must be completed before reduce task can start processing. To finish the job before all map tasks are completed and if we get the desired results, then the following tools may be useful:

    • Sampling: EARL
    • Sorting: RanKloud

4. Lack of iteration: MR programmers need to write a sequence of MR jobs and coordinate their execution to implement machine learning algorithms. The intermediate result of each iteration is stored in HDFS and then reloaded for further processing. It involves huge IO-bound processing. Possible tools to optimize these are:

    • Looping, caching, pipelining: Stratosphere, Haloop, NOVA, Twister, CBP, Pregel.
    • Incremental processing: Stratosphere, REX, Differential dataflow

5. Lack of interactive and stream processing: Fault-tolerance in MR leads to various overheads that impact the performance of a job. Many applications require fast response time for interactive and stream analytics. Possible tools are to facilitate this are:

   - Streaming, pipelining: Dremel, Impala, Hyracks, Tenzing
   - In-memory processing: PowerDrill, Spark/Shark, M3R
   - Pre-computation: BlikDB

6. Joins are slow in MR.
7. There is no specific optimal configuration for job efficiency.

KEYWORDS

- **failure handling in HDFS and MapReduce**
- **Hadoop MapReduce weaknesses and solutions**
- **HDFS and MapReduce components**
- **MapReduce execution flow**
- **MapReduce phases**

CHAPTER 3

Hadoop 1.2.1 Installation

We generate fears while we sit. We overcome them by action.

—Henry Link

INTRODUCTION

Hadoop is built to natively work with Linux. If the default OS is not Linux, then use hypervisor (Virtual box or KVM) to create a Linux VM and try Hadoop. However, Hadoop works with Windows, MAC, Solaris, etc. OSs as well. MR (JT, TT) and HDFS (NN, SNN, DN) components run as daemons in Physical Machines (PM) or VMs. PMs and VMs are denoted as nodes in general. A daemon is a background process that keeps running. There are different modes of Hadoop deployment.

Single node implementation: Hadoop in one machine

1. Standalone or Local mode: No Hadoop daemons run in this mode. All execution sequence is taken care of by the Hadoop framework itself. It is useful to test and debug MR jobs before launching in a large Hadoop cluster.
2. Pseudo-distributed mode: MR and HDFS daemons run in separate JVM by simulating cluster environment within a node by using more than one JVM.

Multi-node implementation: Hadoop in more than one node

1. Fully distributed or cluster mode: MR and HDFS daemons run in different PMs in the cluster.
2. Virtual cluster mode: MR and HDFS daemons run in different VMs in a virtual cluster.

3.1   SYSTEM REQUIREMENTS

I am going to use a PM installed with Linux flavor OS (Ubuntu 18.04). If you do not use Linux, install virtual box (or any hypervisor) and create a VM with Ubuntu 18.04. To check Ubuntu version

```
$ lsb_release -a
   Distributor ID:      Ubuntu
   Description:         Ubuntu 18.04.1 LTS
   Release:             18.04
   Codename:            bionic
```

Create a new username (optional), and add it to sudoers list for full privilege.

```
$ sudo adduser itadmin
$ sudo adduser itadmin sudo
```

Login to itadmin and follow the steps given below to prepare this node suitable for Hadoop installation.

1. Update and upgrade the node for supporting the latest packages.

```
$ sudo apt update
$ sudo apt upgrade
```

2. Install vim for editing text files more comfortably.

```
$ sudo apt install vim -y
```

3. Install Openssh for remote access to Linux nodes and communication among Hadoop nodes.

```
$ sudo apt install openssh-server
```

4. If dynamic IP is enabled for Hadoop nodes, then IP may change intermittently, which will disturb the Hadoop cluster. Therefore, it is necessary to set static IP and verify the connection. If you use VM, the verify the connection between host to Internet, host to VM, VM to Internet, and VM to host.

   We need to check the available network interfaces in a node. Available network interfaces in my machine are given below for an example. This might be different in your machine.

```
$ ifconfig -a
eth0: flags=4163<UP,BROADCAST,RUNNING,MULTI
      CAST> mtu 1500 inet 10.100.52.61 netmask
      255.255.0.0 broadcast 10.100.255.255
```

```
inet6 fe80::8eec:4bff:fe7d:8242 prefixlen 0x20<link>
ether 8c:ec:4b:7d:82:42 txqueuelen 1000 (Ethernet)
lo: flags=73<UP,LOOPBACK,RUNNING> mtu 65536
    inet 127.0.0.1 netmask 255.0.0.0 inet6
    ::1 prefixlen 128 scopeid 0x10<host> loop
    txqueuelen 1000 (Local Loopback)
```

I set static IP with 10.100.55.92 for my node and provide netmask, gateway, and nameservers according to my local network configuration. This information could be different in your working environment. So, be careful while entering this information.

$ sudo vi /etc/netplan/01-netcfg.yaml

```
network:
  version: 2
  renderer: networkd
  ethernets:
    eth0:
      dhcp4: no
      dhcp6: no
      addresses: [10.100.55.92/16]
      gateway4: 10.100.52.1
      nameservers:
      addresses: [10.20.1.22,8.8.8.8,8.8.4.4]
```

$ sudo netplan apply   // restart the network interface
$ sudo netplan —debug apply // gives you the error details if any

Caution:

Be careful with spacing between key-value and in other places. If it is not proper, then static IP will not be set.

10.100.55.92/16 is a CIDR notation. I have used 16 because my local network netmask is 255.255.0.0. You can find this in the following link.

```
CIDR notation for 255.255.255.0 is 24
CIDR notation for 255.255.0.0 is 16
```

You can refer https://doc.m0n0.ch/quickstartpc/intro-CIDR.html for more.

5. Disable firewall to allow remote calls from Hadoop daemons.

    $ sudo ufw disable
    $ sudo ufw status

    If you install Hadoop in VM, you have to disable the firewall in both VM and host OS as well.

6. Install the latest version of java

    If you already had installed java, please remove completely to install new Java version. Then, set up path for java in the .bashrc file.

    $ sudo apt install openjdk-8* -y   // installs latest opensource JDK
    $ ls /usr/lib/jvm/

    ```
    java-1.8.0-openjdk-amd64 java-8-openjdk-amd64
    openjdk-8
    ```

    $ vi .bashrc   // to set path to java, append to the end of file

    ```
    export JAVA_HOME=/usr/lib/jvm/
    java-8-openjdk-amd64
    export PATH=$PATH:$JAVA_HOME/bin
    ```

    $ source .bashrc   // to refresh environment file
    $ java -version   // verify java and javac version

    ```
    openjdk version "1.8.0_212"
    OpenJDK Runtime Environment (build
    1.8.0_212-8u212-b03-0ubuntu1.18.04.1-b03)
    OpenJDK 64-Bit Server VM (build 25.212-b03,
    mixed mode)
    ```

    $ javac -version

    ```
    javac 1.8.0_212
    ```

If you use Ubuntu VM on Windows (host OS), you can use:

- WinSCP to transfer files between Windows (host OS) and Ubuntu VM. WinSCP is a file transfer protocol with a nice GUI and installed in Windows.
- PUTTY to remotely log in from Windows (host OS) to Ubuntu VM to launch commands.

We are going to use opensource Hadoop software from Apache BigTop project for Hadoop installation. Therefore, download a stable release, which is packaged as a gzipped tar file from the Apache Hadoop release page.

If you want to install Hadoop in Windows machines, install Cygwin and SSH server. It provides a Unix-like environment on Windows. The following link provides step-by-step instructions. http://opensourceforu.com/2015/03/getting-started-with-hadoop-on-windows/

3.2 SINGLE NODE SETUP

3.2.1 *STANDALONE OR LOCAL MODE*

By default, Hadoop has been configured to run in standalone mode as a single java process, as shown in Figure 3.1. In local mode, no Hadoop daemons are running. So, there is no concept of HDFS and MR in standalone implementation. It works on a local file system with the default configuration. We need not configure any files as part of the installation. All execution sequence is handled by the Hadoop framework itself. This is recommended only for practicing Hadoop commands and debug MR jobs before launching in a real Hadoop cluster.

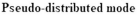

No Hadoop daemons	NN	SNN	JT	TT	DN
JVM	JVM	JVM	JVM	JVM	JVM
OS	OS
Hardware	Hardware

Local or standalone mode   Pseudo-distributed mode

FIGURE 3.1 Single node setup.

Step 1: Download Hadoop 1.2.1 and untar

- If Hadoop file is in Windows, use WinSCP to copy from Windows to Ubuntu or
- You can download Hadoop tarball release from apache software foundation. https://archive.apache.org/dist/hadoop/core/hadoop-1.2.1.tar.gz
- Else, download using command
  $ wget https://archive.apache.org/dist/hadoop/core/hadoop-1.2.1/hadoop- 1.2.1.tar.gz
- Untar Hadoop file
  $ tar -zxvf hadoop-1.2.1.tar.gz

Step 2: Move Hadoop file to /usr/local/hadoop

$ sudo cp -r hadoop-1.2.1 /usr/local/hadoop

Step 3: Set path to Hadoop

$ vi ~/.bashrc   // append to the end of file

```
export HADOOP_HOME=/usr/local/hadoop
export PATH=$PATH:$HADOOP_HOME/bin
```

$ source .bashrc

$ $PATH   // verify the path (text in bold)

```
-bash: /home/itadmin/bin:/home/itadmin/.local/bin:/
usr/local/sbin:/usr/local/bin:/usr/sbin:/usr/bin:/
sbin:/bin:/usr/games:/usr/local/games:/snap/bin:/
```
`usr/local/hadoop/bin:/usr/lib/jvm/java-8-openjdk-`
`amd64/bin:``No such file or directory`

Step 4: Configure Hadoop environment variable

You do not have to edit any configuration files except hadoop-env.sh to set JAVA_HOME and enabling IPV4.

$ sudo vi /usr/local/hadoop/conf/hadoop-env.sh   // append to the end of file

```
  export JAVA_HOME=/usr/lib/jvm/java-8-openjdk-amd64
  export HADOOP_OPTS=-Djava.net.preferIPv4Stack=true
```

$ hadoop version   // Hadoop version is displayed

```
  Warning: $HADOOP_HOME is deprecated.
  Hadoop 1.2.1
  Subversion https://svn.apache.org/...branch-1.2
  -r 1503152
  Compiled by mattf on Mon Jul 22 15:23:09 PDT 2013
  From source with checksum 6923c86528809c4e7e6f493
  b6b413a9a
  This command run using /usr/local/hadoop/
  hadoop-core-1.2.1.jar
```

Step 5: Give privileges

Grant all access rights to the Hadoop daemons to read and write from a local file system location /usr/local/hadoop. Username is **itadmin.**

```
$ sudo chown -R itadmin /usr/local/hadoop
$ sudo chmod -R 777 /usr/local/hadoop
```

Step 6: Run a simple wordcount job

There are no MR and HDFS daemons running in this mode. Create a file "input.txt" in the local file system and launch MR wordcount job with output directory "result," which is created in the local file system, not in HDFS. Output directory should not already exist. Once the job is submitted, Local-JobRunner will take care of the MR job execution sequence as there is no MR daemon running.

```
$ vi input.txt
      hi how are you
```

Launch MR wordcount job. The syntax is,

```
$ hadoop jar jar_location job_name input_file output_directory_name
$ hadoop jar /usr/local/hadoop/hadoop-examples-1.2.1.jar wordcount
input.txt result
$ ls result
```

`part-r-00000`	// output file
`_SUCCESS`	// empty file showing MR execution success

`$ cat result/part-r-00000` // to view the wordcount output

```
are     1
hi      1
how     1
you     1
```

Give a different name for output directory for every run. You do not have any WUI facility in this mode as no configurations are set.

3.2.2   PSEUDO-DISTRIBUTED MODE

Each Hadoop daemon runs in different JVM in a node, as shown in Figure 3.1. As HDFS and MR daemons are running in different JVM, it seems like they are virtually running in different nodes thus, it gets the name "pseudo-distributed mode." Please, refer Section 3.1 to complete the system requirements. Table 3.1 shows the configuration files we are going to edit.

TABLE 3.1   HDFS and MR Configuration Files

Configuration file	Description
hadoop-env.sh	To specify the location of java and configure HDFS and MR daemons.
core-site.xml	To specify NN, location for storing FSImage in NN and SNN, and location to store blocks in DNs.
mapred-site.xml	To configure MR job execution sequence.
hdfs-site.xml	To configure NN, SNN, and DN.
Masters	To specify SNN for checkpointing.
Slaves	To set a list of slave nodes (one per line) that runs DN and TT.

3.2.2.1   INSTALLATION IN SHORT

1. Generate public-private key pair for passwordless communication.
2. Download Hadoop 1.2.1 and untar.
3. Move Hadoop file to /usr/local/hadoop location.
4. Set path to Hadoop in .bashrc file.
5. Configure Java_Home, disable IPv6 in hadoop-env.sh
6. Edit configuration files as given in Table 3.1.
7. Change permission and ownership for Hadoop location in the local file system.
8. Create HDFS using format command (creates a namespace for NN).
9. Launch Hadoop daemons.

3.2.2.2   INSTALLATION IN DETAIL

Step 1: Generate public-private key pair for passwordless communication

Hadoop daemons rely on SSH protocol to communicate with each other and perform cluster-wide operations. So, a public-private key pair is generated and added to the authorized_keys file for passwordless communication to start/stop services. Otherwise, every command to start/stop services will prompt username and password.

```
$ ssh localhost   // asks username and password and loops into same host
$ exit            // to exit from loop
```

Generate and add public key to ~/.ssh/authorized_keys file. Once added, you can log into your system itself without password. In a production environment, passphrase is for security purpose.

```
$ ssh-keygen or ssh-keygen -t rsa   // leave empty for name, password
$ ls ~/.ssh                          // you can see the public-private keys
$ vi ~/.ssh/id_rsa.pub               // to view public key

$ cat ~/.ssh/id_rsa.pub >> ~/.ssh/authorized_keys

$ chmod 0700 ~/.ssh/authorized_keys  // owner can read, write and
                                        execute
$ ssh localhost                      // prompts no username and
                                        password
$ exit                               // to log out from loop
```

Step 2: Download Hadoop 1.2.1 and untar

- If you have Hadoop file in Windows, use WinSCP to copy from Windows to ubuntu or
- You can download Hadoop tarball release from apache software foundation.
  https://archive.apache.org/dist/hadoop/core/hadoop-1.2.1.tar.gz
- Else, download using command
  $ wget https://archive.apache.org/dist/hadoop/core/hadoop-1.2.1/hadoop-1.2.1.tar.gz
- Untar the Hadoop file
  $ tar -zxvf hadoop-1.2.1.tar.gz

Step 3: Move Hadoop file to /usr/local/hadoop

```
$ sudo cp -r hadoop-1.2.1 /usr/local/hadoop
$ ls /usr/local/hadoop
  bin hadoop-ant-1.2.1.jar   ivy    sbin
  build.xml hadoop-client-1.2.1.jar ivy.xml share
  c++ hadoop-core-1.2.1.jar lib src
  CHANGES.txt hadoop-examples-1.2.1.jar libexec webapps
  conf hadoop-minicluster-1.2.1.jar LICENSE.txt
  contrib hadoop-test-1.2.1.jar NOTICE.txt
  docs hadoop-tools-1.2.1.jar README.txt
```

/usr/local/hadoop/bin – contains script files to start/stop Hadoop daemons.
/usr/local/hadoop/conf – contains configuration files for Hadoop daemons.
/usr/local/hadoop/logs – state of Hadoop activities is recorded. You can also see system log file /var/log if any error.
/usr/local/hadoop/sbin – supporting script files for Hadoop services.

Step 4: Set path to Hadoop in .bashrc file.

$ vi .bashrc   // append to the end of file
```
export HADOOP_HOME=/usr/local/hadoop
export PATH=$PATH:$HADOOP_HOME/bin
```
$ source .bashrc

$ $PATH   // to verify hadoop path setup

Step 5: Configure Java_Home, disable IPv6 in hadoop-env.sh

$ sudo vi /usr/local/hadoop/conf/hadoop-env.sh   // add into the file
```
export JAVA_HOME=/usr/lib/jvm/java-8-openjdk-amd64
export HADOOP_OPTS=-Djava.net.preferIPv4Stack=true
```
$ hadoop version   // to verify Hadoop version

Step 6: Edit configuration files

IP and username of my node is 10.100.55.92 and itadmin. Configuration is set using property with name-value pair. Example:

```
name – fs.default.name
value – hdfs://10.100.55.92:10001
```

core-site.xml

$ sudo vi /usr/local/hadoop/conf/core-site.xml
```
<configuration>
  <property>
    <name>fs.default.name</name>
    <value>hdfs://10.100.55.92:10001</value>
  </property>
  <property>
    <name>hadoop.tmp.dir</name>
    <value>/usr/local/hadoop/tmp</value>
  </property>
</configuration>
```

fs.default.name – specifies the IP address of NN.
hadoop.tmp.dir – specifies the local file system location for storing FSImage and edit logs in the NN and SNN, meta-data of JT, to store blocks in DNs.

mapred-site.xml

$ sudo vi /usr/local/hadoop/conf/mapred-site.xml

```
<configuration>
    <property>
        <name>mapred.job.tracker</name>
        <value>10.100.55.92:10002</value>
    </property>
</configuration>
```

mapred.job.tracker – specifies the IP address of the node running JT

masters – to specify the IP of SNN for checkpointing

$ sudo vi /usr/local/hadoop/conf/masters

```
    10.100.55.92
```

slaves – to specify a list of slaves to run TT and DN

$ sudo vi /usr/local/hadoop/conf/slaves

```
    10.100.55.92
```

HDFS configuration – to configure NN, SNN, DN. It is optional.

$ sudo vi /usr/local/hadoop/conf/hdfs-site.xml

```
    <configuration>
        <property>
            <name>dfs.data.dir</name>
            <value>/usr/local/hadoop/tmp/dfs/data</value>
        </property>
        <property>
            <name>dfs.name.dir</name>
            <value>/usr/local/hadoop/tmp/dfs/name</value>
        </property>
        <property>
            <name>dfs.replication</name>
            <value>1</value>
        </property>
        <property>
            <name>dfs.block.size</name>
            <value>64000000</value>
```

```
        </property>
    </configuration>
```

---

hadoop.tmp.dir – specifies the location in NN and SNN to store FSImage, editlogs, and location in DNs to store data blocks. The value for this property can be a single path (/usr/local/hadoop/tmp) or comma separated paths (/disk1/usr/local/hadoop/tmp, /disk2/usr/local/hadoop/tmp) if more than one disk available.

hdfs-site.xml is an optional configuration file. If you do not specify hdoop.tmp.dir property in core-site.xml, then use hdfs-site.xml to specify the location separately /usr/local/hadoop/tmp/dfs/.. for NN, DN, SNN and /usr/local/hadoop/tmp/mapred/.. for JT, TT.

dfs.data.dir – value is /usr/local/hadoop/tmp/dfs/data – DN stores blocks. There can be more than one path specified.

dfs.name.dir – value is usr/local/hadoop/tmp/dfs/name – NN maintains its FSImage, editlogs.

fs.checkpoint.dir – value is /usr/local/hadoop/tmp/dfs/namesecondary – SNN stores checkpointed FSImage, and editlogs.

Now, assume you didn't specify /usr/local/hadoop/tmp in core-site.xml and hdfs-site.xml. Still, you can launch the daemons and perform operations. meta-data directories are created in default locations. However, it is always recommended to set path in core-site. xml or in hdfs-site.xml.

Step 7: Grant access to Hadoop daemons

Grant all access rights to Hadoop daemons to read and write from the local file system location /usr/local/hadoop. **itadmin** is the username.

```
$ sudo chown -R itadmin /usr/local/hadoop/
$ sudo chmod -R 777 /usr/local/hadoop/
```

Step 8: Create HDFS namespace

Execute format command to create a new HDFS namespace with new namespace ID.

```
$ hadoop namenode -format
```

```
......
Storage directory /usr/local/hadoop/tmp/dfs/name
has been successfully formatted.
```

You will see various metrics and a message similar to the one above. This process creates an empty HDFS with storage directories and an initial version of the meta-data in the NN. The DNs are not involved in the initial formatting process.

$ ls /usr/local/hadoop/tmp

```
dfs
```

Step 9: Launch Hadoop daemons: There are many ways to start/stop HDFS and MR services

Way 1: Starting HDFS and MR services separately

$ start-dfs.sh	// NN, SNN, DN are started
$ ls /usr/local/hadoop/tmp/dfs	// directories for SNN and DN are created
$ jps	// JVM process status to see a list of running JVM

```
DataNode
NameNode
SecondaryNameNode
```

jps displays class names of services

$ start-mapred.sh	// JT and TT are started
$ ls /usr/local/hadoop/tmp	// mapred directory is created for JT and TT
$ ls /usr/local/hadoop/tmp/mapred	
$ jps	// you will see JT and TT services running

```
TaskTracker
DataNode
NameNode
JobTracker
SecondaryNameNode
```

All services are running in the same node in single node implementation

Way 2: Starting HDFS and MR together

$ start-all.sh	// to start HDFS and MR services together

```
$ jps
$ stop-all.sh              // to stop HDFS and MR services together
$ jps
```

Way 3: You can start/stop each and every service one by one

```
$ hadoop-daemon.sh start/stop namenode
$ hadoop-daemon.sh start/stop secondarynamenode
$ hadoop-daemon.sh start/stop datanode
$ hadoop-daemon.sh start/stop jobtracker
$ hadoop-daemon.sh start/stop tasktracker
$ jps
```

3.2.2.3   MR WORDCOUNT JOB

```
$ vi input.txt              // enter some text
    hi how are you
$ hadoop fs -ls /           // lists meta-data of files from HDFS root
$ hadoop fs -mkdir /data    // create a directory in HDFS
$ hadoop fs -copyFromLocal input.txt /data   // load input.txt onto /data
                                                                directory
$ hadoop fs -ls /data
$ hadoop fs -cat /data/input.txt        // displays meta-data
$ hadoop jar /usr/local/hadoop/hadoop-examples-1.2.1.jar wordcount
/data/input.txt /result
```

- result is a directory name and should not already exist in HDFS. Give different name for output directory for every run.
- wordcount is a job (class name) that contains the main method to start MR job in hadoop-examples-1.2.1.jar.

3.2.2.4   VALIDATING OUTPUT

Output and other error messages are displayed in console and recorded in log files for later usage. There are different ways to display MR job results: HDFS commands, Web User Interface, HDFS API in java programs.

HDFS commands

```
$ hadoop fs -ls /result
Found 3 items
-rw-r--r-- 3 rathinaraja supergroup 0 2017-02-15
20:54 /result/_SUCCESS
drwxr-xr-x - rathinaraja supergroup0 2017-02-15
20:54 /result/_logs
-rw-r--r-- 3 rathinaraja supergroup 13 2017-02-15
20:54 /result/part-r-00000
```

`_SUCCESS` – It is an empty file to indicate job success
`_logs` – It is a directory that contains job history
`part-r-00000` – It is the output file

```
$ hadoop fs -cat /result/part-r-00000        // to view wordcount output
are      1
hi       1
how      1
you      1
```

You can bring the result to local file system and then view using Linux commnds.

```
$ hadoop fs -copyToLocal /result ~/
$ cat result/part-r-00000
```

Web User Interface (WUI)

To see the result via WUI, open browser and enter IP address + port number of the node where Hadoop services are running. In single node implementation, HDFS and MR services are running in the same node.

IP of NN:50070   // after start-dfs.sh you can see data.

IP of DN:50075   // to see DN

IP of SNN:50090   // to see SNN

IP of JT:50030   // after start-mapred.sh you can get jobs execution details

IP of TT:50060   // to see what is going on in TT

To check the result, enter NN_IP:50070 in the web browser as shown in Figure 3.2, browse the files for result directory. There you will see "part-r-00000" file. You can also explore the live DNs contributing its storage to Hadoop cluster, HDFS capacity used/unused, dead DNs, etc.

10.100.55.92:50070/

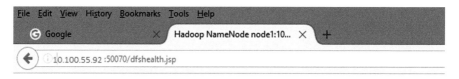

NameNode 'node1:10001'

Started:	Sat Sep 24 16:41:12 IST 2016
Version:	1.2.1, r1503152
Compiled:	Mon Jul 22 15:23:09 PDT 2013 by mattf
Upgrades:	There are no upgrades in progress.

Browse the filesystem
Namenode Logs

Cluster Summary

20 files and directories, 11 blocks = 31 total. Heap Size is 60 MB / 889 MB (6%)

Configured Capacity	:	5.11 GB
DFS Used	:	101.64 KB
Non DFS Used	:	4.52 GB
DFS Remaining	:	605 MB
DFS Used%	:	0 %
DFS Remaining%	:	11.57 %
Live Nodes	:	1
Dead Nodes	:	0
Decommissioning Nodes	:	0
Number of Under-Replicated Blocks	:	5

NameNode Storage:

Storage Directory	Type	State
/usr/local/hadoop/tmp/dfs/name	IMAGE_AND_EDITS	Active

FIGURE 3.2   Namenode WUI.

To see job status, enter JT_IP:50030 in the web browser, as shown in Figure 3.3 and browse the completed jobs. The number of map/reduce task launched and their attempts, number of bytes read, number of bytes written, etc. are displayed. All these parameters are a job and task counters. Moreover, you can explore the resource capacity available, details of TTs, number of map/reduce slots, etc.

10.100.55.92:50030/

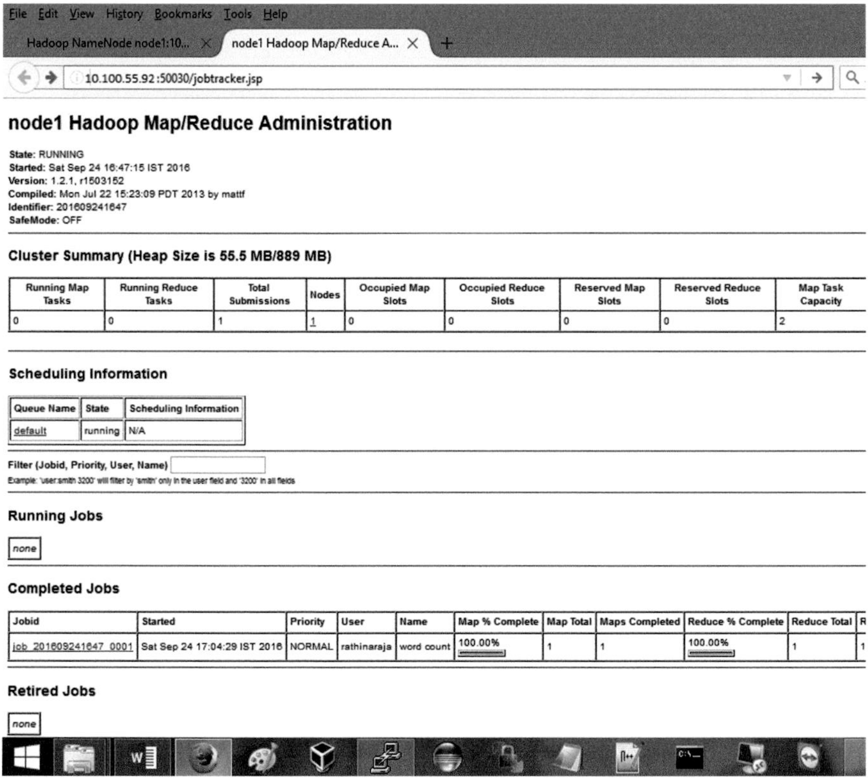

FIGURE 3.3 MapReduce WUI.

HDFS API in java programs

We can use HDFS APIs in java program to interact with running Hadoop daemons and display information. To achieve this, the java program should use HDFS URL to connect with running daemons.

Once the HDFS daemons are stopped, they can be restarted again. If you format NN again, new namespaceID is created. So, the DNs having old namespaceID will not match with the new namespace version leading to cluster down. So, if there is any problem in starting NN, first delete /usr/local/hadoop/tmp directory and re-format the NN. Consequently, you will lose all old data and meta-data form HDFS.

$ sudo rm -r /usr/local/hadoop/tmp
$ hadoop namenode -format

To practice from the beginning, remove Hadoop file from /usr/local and follow from Step 1

$ sudo rm -r /usr/local/hadoop

3.3   MULTI-NODE SETUP: fully distributed mode and virtual cluster mode deployment

All master processes (NN, JT, SNN) is deployed in the same node only for testing and training purposes. For a smaller production deployment, at least the SNN should be on a different node from the NN. The fully distributed mode can be established on either cluster of PMs, as shown in Figure 3.4 or cluster of VMs, as shown in Figure 3.5. A virtual cluster contains a set of VMs running Hadoop services (shaded VMs). For large production deployment, NN, JT, and SNN should be deployed on a dedicated node.

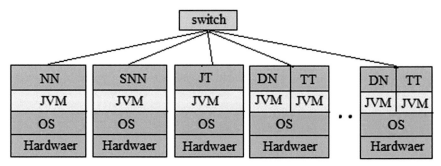

Fully distribtued or Cluster mode

FIGURE 3.4   Fully distributed or cluster mode on PMs.

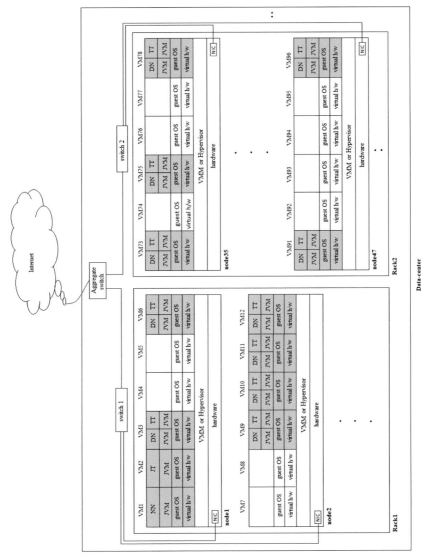

FIGURE 3.5   Multi-node implementation on cluster of VMs.

Table 3.2 gives you a fair idea about the different size of the cluster and its deployment on the physical/virtual cluster. Table 3.3 accounts typical hardware configuration required to run individual Hadoop components in multi-node setup. Chapter 6 will give you more detail about the Hadoop cluster and node configuration.

TABLE 3.2   Multi-Node Flavor

Small cluster	Medium/single-rack	Large/multi-rack
2–10 nodes	10–40 nodes	100+ nodes
NN, JT, SNN on the same node	NN and JT on the same node	NN, JT, and SNN on dedicated nodes
DN and TT on all other nodes	SNN on a dedicated node DN and TT on all other nodes	TT and DN on all other nodes feed rack awareness to NN and JT perform network, HDFS, MR optimizations

TABLE 3.3   Recommended Multi-Node Server Configuration in Data-Centres

Components	NN and JT	SNN	Slaves (DN+TT)
CPU	Two xeon dual core hyper-threaded	Two xeon dual core	Two xeon quad core hyper-threaded
Memory	8+ GB	8 GB	32 GB
HDD	1 TB	1 TB	4 x 2 TB
Ethernet	1 Gb/s	1 Gb/s	1 Gb/s

How large should a cluster be? There is no one specific answer to that question. It is scalable as much as you need to achieve parallelism. You need to provide rack awareness for NN and JT manually in the multi-rack environment.

3.3.1   OS TYPE AND JAVA VERSION IN ALL PHYSICAL/VIRTUAL NODES

1. OS type must be same in all nodes (either 32-bit OS or 64-bit OS, but not mixed up). Use the following command to check OS type. If it is not the same, it is better to install the right version.

$ uname -a

Result for 32-bit Ubuntu will be as below:

```
Linux discworld 2.6.38-8-generic #42-Ubuntu SMP Mon
Apr 11 03:31:50 UTC 2011 i686 i686 i386 GNU/Linux
```

Result for 64-bit Ubuntu will be as below:

```
Linux ubuntu 4.4.0-104-generic #127-Ubuntu SMP Mon
Dec 11 12:16:42 UTC 2017 x86_64 x86_64 x86_64 GNU/
Linux
```

Alternatively, you can use the following command

$ uname -i

```
x86_64
```

2. Install Java in all the nodes with the same version (OpenJDK/Oracle JDK) and architecture (32-bit/64-bit). Refer Section 3.1 to install Java. To check java version and architecture,

    $ java -version

    ```
    openjdk version "1.8.0_131"
    OpenJDK Runtime Environment (build
    1.8.0_131-8u131-b11-2ubuntu1.16.04.3-b11)
    OpenJDK 64-Bit Server VM (build 25.131-b11, mixed
    mode)
    ```

    $ javac -version

    ```
    javac 1.8.0_131
    ```

    or go to /usr/lib/jvm

    $ ls /usr/lib/jvm

    Result for 32-bit java will be as below:

    ```
    java-1.8.0-openjdk-i386 java-8-openjdk-i386
    ```

    Result for 64-bit Ubuntu:

    ```
    java-1.8.0-openjdk-amd64 java-8-openjdk-amd64
    ```

3.3.2 MULTI-NODE INSTALLATION

The requirements given in Section 3.1 should be met for all nodes in the physical/virtual cluster. Most of us would not be having multiple PMs to experiment fully distributed mode. So, we can create 5 VMs for our experiment. The minimum requirements to launch 5 VMs in a PM are to have a quad-core processor, 16 GB memory, and 500 GB HDD. Values for IP, hostname, and domain name are given in Table 3.4. These values are from my local LAN. You have to fill this table according to your LAN configurations.

The term "node" here means either PM or VM. A domain name can be anything, but I have used a meaningful term "node." If you can launch only two VMs in your machine, then launch NN+JT+SNN in VM1 and DN+TT in VM2. You can also launch multiple VMs in different PMs. Whether a cluster of VMs or PMs, the following procedure works just fine.

Caution:

1. Username must be the same in all the nodes. I have set "**itadmin**" as the username for all the nodes in my cluster. So, create a common username in all nodes in the cluster using the following command.

   ```
   $ sudo adduser itadmin      // leave empty for all the fields
   $ sudo adduser itadmin sudo
   ```

   Now, username itadmin is created. Log into itadmin and do the rest.

2. Another important restriction is, the hostname of the nodes must be unique across the cluster; that is, the hostname should not be identical in the cluster. To change the hostname to ubuntu1 in your machine without restarting, use the following commands

   ```
   $ sudo vi /etc/hostname
       ubuntu1
   ```

   ```
   $ sudo vi /etc/hosts
       127.0.0.1    localhost
       127.0.1.1    ubuntu1
   ```

   ```
   $ sudo sysctl kernel.hostname=ubuntu1
   ```

   Do this in all nodes and create unique hostname as given in Table 3.4.

Step 1: Set the domain name in all the nodes in the cluster

Go to /etc/hosts to set DNS. After setting the domain name, you can use domain names instead of IP address. Comment 127.0.*.1 and enter IP, domain name, and hostname of each node in a line with tab space separator (refer Table 3.4).

TABLE 3.4 Multi-Node Details

VM/PM	Daemon	IP	Hostname	Domain name
node1	NN	10.100.55.92	ubuntu1	node1
node2	JT	10.100.55.93	ubuntu2	node2
node3	SNN	10.100.55.94	ubuntu3	node3
node4	Slave1 (DN+TT)	10.100.55.95	ubuntu4	node4
node5	Slave2 (DN+TT)	10.100.55.96	ubuntu5	node5

$ sudo vi /etc/hosts

```
#127.0.0.1          localhost
#127.0.1.1          ubuntu1
# IP                domain_name          host_name
10.100.55.92        node1                ubuntu1
10.100.55.93        node2                ubuntu2
10.100.55.94        node3                ubuntu3
10.100.55.95        node4                ubuntu4
10.100.55.96        node5                ubuntu5
```

Step 2: Set passwordless communication

Create an SSH key in NN and send to all other nodes. Similarly, create an SSH key in JT and send it to all the other nodes. You need not create any key in slaves and SNN. In case, if you are hosting NN and JT in the same node, it is sufficient to do once.

Create key in node1 (NN) and send to all other nodes in the cluster

```
$ ssh-keygen or ssh-keygen -t rsa
$ cat ~/.ssh/id_rsa.pub >> ~/.ssh/authorized_keys
$ chmod 0700 ~/.ssh/authorized_keys
$ ssh-copy-id -i ~/.ssh/id_rsa.pub username@IP // to send keys to
other nodes
$ ssh-copy-id -i ~/.ssh/id_rsa.pub itadmin@node2
$ ssh-copy-id -i ~/.ssh/id_rsa.pub itadmin@node3
$ ssh-copy-id -i ~/.ssh/id_rsa.pub itadmin@node4
$ ssh-copy-id -i ~/.ssh/id_rsa.pub itadmin@node5
```

Try to remotely login from node1 to all other nodes in the cluster using ssh command without password. For example, from node1

```
$ ssh node2
$ exit
```

Similarly, create a ssh key in node2 (JT), send it to all the other nodes (1,3,4,5), and verify the passwordless connection.

Step 3 to Step 8 must be done in all nodes in the cluster

Hadoop does not have a single, global location for configuration information. Instead, each node in the cluster has its configuration files, and it is up to the administrator to ensure that they are kept in sync at any point in time.

Step 3: Download Hadoop and untar

- If you have Hadoop file in Windows, use WinSCP to copy from Windows to ubuntu or
- You can download Hadoop tarball release from apache software foundation.
  https://archive.apache.org/dist/hadoop/core/hadoop-1.2.1.tar.gz
- Else, download using command
  $ wget https://archive.apache.org/dist/hadoop/core/hadoop-1.2.1/hadoop-1.2.1.tar.gz
- Untar the Hadoop file
  $ tar -zxvf hadoop-1.2.1.tar.gz

Step 4: Move Hadoop file to /usr/local/hadoop

```
$ sudo cp -r hadoop-1.2.1 /usr/local/hadoop
$ ls /usr/local/hadoop
```

Step 5: Set path to Hadoop

```
$ vi ~/.bashrc
    export HADOOP_HOME=/usr/local/hadoop
    export PATH=$PATH:$HADOOP_HOME/bin
$ source .bashrc
$ hadoop version
$ $PATH
```

Step 6: Configure Java_Home, disable IPv6 in hadoop-env.sh

```
$ sudo vi /usr/local/hadoop/conf/hadoop-env.sh
    export JAVA_HOME=/usr/lib/jvm/java-8-openjdk-amd64
```

```
export HADOOP_OPTS=-Djava.net.preferIPv4Stack=true
```
$ hadoop version

Step 7: Edit configuration files

core-site.xml – to specify NN IP address

$ sudo vi /usr/local/hadoop/conf/core-site.xml

```
<configuration>
        <property>
                <name>fs.default.name</name>
                <value>hdfs://node1:10001</value>
        </property>
        <property>
                <name>hadoop.tmp.dir</name>
                <value>/usr/local/hadoop/tmp</value>
        </property>
</configuration>
```

mapped-site.xml – to specify JT IP address

$ sudo vi /usr/local/hadoop/conf/mapred-site.xml

```
<configuration>
        <property>
                <name>mapred.job.tracker</name>
                <value>node2:10002</value>
        </property>
</configuration>
```

hdfs-site.xml – to configure HDFS components

$ sudo vi /usr/local/hadoop/conf/hdfs-site.xml

```
<configuration>
        <property>
                <name>dfs.replication</name>
                <value>3</value>
        </property>
</configuration>
```

masters – to specify SNN IP address

```
$ sudo vi /usr/local/hadoop/conf/masters
    node3
```

slaves – to specify slave nodes that run TT and DN together

```
$ sudo vi /usr/local/hadoop/conf/slaves
    node4
    node5
```

Step 8: Change ownership and grant access to Hadoop services

```
$ sudo chown -R itadmin /usr/local/hadoop
$ sudo chmod -R 777 /usr/local/hadoop
```

Step 9: Create HDFS namespace (in node1)

Once all nodes are configured successfully, then format HDFS to create namespace.

```
$ hadoop namenode -format
```

/usr/local/hadoop/tmp directory is created after this command in node1 (NN)

```
$ ls /usr/local/hadoop/tmp
    dfs
$ ls /usr/local/hadoop/tmp/dfs
    name
$ ls /usr/local/Hadoop/tmp/dfs/name
    current        image
$ ls /usr/local/Hadoop/tmp/dfs/name/current
  edits    fsimage    fstime    VERSION
```

Step 10: Launch Hadoop daemons

Start the HDFS first using the following command in node1 (NN). It starts the NN in node1, SNN in node3 and DN in node4, node5.

```
$ start-dfs.sh

$ jps          // run this command in NN, SNN, DN to ensure appropriate
                  Hadoop services are running.
```

> in SNN (node3), namesecondary directory is created
>
> $ ls /usr/local/hadoop/tmp/dfs
> ```
> namesecondary
> ```
> in slaves (node3, node4), data directory is created
>
> $ ls /usr/local/hadoop/tmp/dfs
> ```
> data
> ```
> $ ls /usr/local/hadoop/tmp/dfs/data/
> ```
> blocksBeingWritten current detach in_
> use.lock storage tmp
> ```

Start MR in node2 (JT). It starts JT in node2 and TT in node4, node5.

```
$ start-mapred.sh
$ jps        // in JT and TT to ensure services running.
```

> In JT (node2), mapred directory is created in /usr/local/hadoop/tmp/ for maintaining job-related meta-data information. Similarly, in slaves (node3, node4) also, mapred directory is created.
>
> $ ls /usr/local/hadoop/tmp
> ```
> mapred
> ```
> $ ls /usr/local/hadoop/tmp/mapred
> ```
> local
> ```
> $ ls /usr/local/hadoop/tmp/mapred/local
> ```
> taskTracker tt_log_tmp ttprivate userlogs
> ```

To stop services, follow the order

```
$ stop-mapred.sh     // MR should be stopped first in node2
$ stop-dfs.sh        // then HDFS is stopped next in node1
```

Commissioning/decommissioning (scalability) nodes from the multi-node cluster are discussed in Section 6.3. Please refer to APPENDIX A and Ref. [18] for big data sources. Refer Section 3.2.2.3 and Section 3.2.2.4 to run simple wordcount job and validate the output using WUI. Launching commands for HDFS and MR can be done from any node in the cluster as all nodes have the same configuration of master and slaves. Ultimately, you will get the same result. However, test with a small and large dataset to observe the running time that Hadoop takes. Hadoop takes more time for small files than processing with the local file system. Because there is a lot of intermediate steps involved, which takes a considerable amount of time. However, with the large dataset, you can observe Hadoop outperforming local file system processing. Therefore, upload GB size of data in the multi-node cluster and then launch MR wordcount job.

If the job produces huge output, then there will create many output files (each is an HDFS block) in the output directory. We have to open the file one by one to view the output. However, there is a way to view the entire output files as a single file.

```
$ hadoop fs -cat /output/part-*        // or bring them to local file system
$ hadoop fs -getmerge /output_dir_in_hdfs file_name_local
$ hadoop fs -getmerge /output result.txt
$ cat result.txt
```

You are not constrained to deploy HDFS and MR daemons in any different combinations as given in Table 3.5 depending upon the number of nodes available in the cluster.

TABLE 3.5   Different Choice of Multi-Node Implementation

Choices	node1	node2	node3	node4	node5
1	NN, SNN, JT, DN, TT	-	-	-	-
2	NN, SNN, JT	DN, TT	DN, TT	DN, TT	DN, TT
3	NN, SNN	JT	DN, TT	DN, TT	DN, TT
4	NN	SNN	JT	DN, TT	DN, TT
5	NN, DN, TT	SNN, DN, TT	JT, DN, TT	DN, TT	DN, TT

In order to experiment with huge data, you can download over 500 GB size of data from [21]. If it is not possible to download, then you can also generate a huge random text file with the size you wanted using the following command. To generate 5 GB text file, specify the size in bytes.

```
$ yarn jar /usr/local/hadoop/share/hadoop/mapreduce/hadoop-
mapreduce-examples-2.7.0.jar teragen 5000000000 /terasort-input
```

3.3.3 *FREQUENTLY USED HADOOP COMMANDS*

Hadoop commands are very similar to Linux commands. You need to remotely log in to any one of the nodes in the Hadoop cluster to issue commands.

$ hadoop – lists all possible subcommands such as namenode, second-arynamenode, datanode, namenode -format, dfsadmin, mradmin, fsck, fs, jobtracker, tasktracker, historyserver, balancer, job, queue, version, jar.

$ hadoop fs – to interact with HDFS. Its sub commands are: put, remove, cat, ls, copy to local, move to local, mkdir, chown.

$ hadoop fsck – to query about file system-related information like BRs.

$ hadoop dfsadmin – the HDFS cluster administrative commands are used to get statistics, refresh nodes, set balancer bandwidth, etc.

$ hadoop job – helps to get job-related information and see/kill running jobs

$ hadoop jar – to launch a job.

$ hadoop jobtracker – interacts with JT node.

$ hadoop namenode – interacts with NN.

$ hadoop datanode – interacts with DN.

HDFS and MR commands

File system shell is an interface between user and HDFS.

Uniform Resource Identifiers (URI) are used to locate resources like files. All the file system shell commands take URIs as arguments. Example:

HDFS: hdfs://NN_IP:port#/file/location

Hadoop Archive (HAR): har:///location/file.har

Local file system: file:///location/file/name

Every command returns exit code either 0 for success or -1 for error.

```
$ hadoop fs                    // fs interacts with HDFS
$ hadoop fs -help              // lists short summary of all commands
$ hadoop fs -help commandName  // displays short summary of commands
$ hadoop fs -help put
```

1. To list meta-data from HDFS root (/) and create directory in HDFS

```
$ hadoop fs -mkdir /dir1    // to create directory in HDFS root
$ hadoop fs -ls /           // to list meta-data
```

```
/usr/local/hadoop/tmp/
/dir1
```

$ hadoop fs -mkdir dir2   //directory is created in HDFS username
$ hadoop fs -ls /   //displays meta-data of hadoop root dir
```
/dir1
/user/itadmin/dir2
/usr/local/hadoop/tmp/
```

$ hadoop fs -ls   // by default HDFS stores in home
```
/user/itadmin/dir2
```
$ hadoop fs -ls hdfs://NN_IP:10001/ // in case of multi node
$ hadoop fs -ls -R /   // to show all hidden files and sub-dir
$ hdoop fs -ls file:///   // lists out local root file system

2. To copy/move files from local file system to HDFS

    $ hadoop fs -put local_source /hdfs_destination
    $ hadoop fs -copyFromLocal local_source /hdfs_destination
    $ hadoop fs -moveFromLocal local_source /hdfs_destination

    If the source is a file, destination can be a file or directory

    If the source is a directory, destination must be a directory

    put vs copyFromLocal: put copies more than one files at the time and
    read directly from stdin.

3. To copy/move files from HDFS to local file system

    $ hadoop fs -get /hdfs_source local_destination
    $ hadoop fs -copyToLocal /hdfs_source local_destination
    $ hadoop fs -moveToLocal /hdfs_source local_destination

4. To display a file from HDFS

    $ hadoop fs -cat /filename

5. To delete a file/directory from HDFS

    $ hadoop fs -rm /filename
    $ hadoop fs -rmr /directory

6. To copy and move a file from one location to another location in HDFS
    itself-

    $ hadoop fs -cp /hdfs_source /hdfs_destination
    $ hadoop fs -mv /hdfs_source /hdfs_destination

7. Other file system operations

   $ hadoop dfsadmin -report   // displays BR

   $ hadoop fsck /hdfs_directory -files -blocks -locations
   $ hadoop fs -du /<file/directory name>   // displays the size of files
                                                        in bytes
   $ hadoop fs -cat /hdfs_file | head -n 5   // displays top five lines
   $ hadoop fs -setrep number /filename   // replication by default 3

8. MapReduce commands

   $ hadoop job -list   // lists currently running applications
   $ hadoop job -kill <jobid>
   $ hadoop job -list all   // returns history of jobs

9. To transfer files from one node to another node, make sure the target location has write permission of its current user.

   $ sudo chown -R itadmin /usr/local/hadoop   // at location of target
                                                          machine
   $ sudo chmod -R 777 /usr/local/hadoop   // at location of target
                                                          machine
   $ scp -r directory username@IP:~/   // this is from our machine

10. Working with jars

    $ hadoop jar /usr/local/hadoop/hadoop-examples-1.2.1.jar
    // lists available jobs in jar
    $ jar tvf name.jar   // lists the contents of jar

You can create an empty file, but cannot edit any files in HDFS. You must create input files in the local file system and move onto HDFS. Therefore, vi, gedit, etc. do not work with HDFS.

   $ hadoop fs -touchz /file.txt
   $ hadoop fs -ls /

Check the following link for more commands https://hadoop.apache.org/docs/r1.2.1/commands_manual.html

3.4 WRITING USER-DEFINED MR JOB/APPLICATION USING ECLIPSE

So far, we have been executing a predefined wordcount job that is part of Hadoop distribution. Now, let us write a wordcount job manually using

Eclipse and run in a Hadoop environment. Good understanding of Java utility and collection packages help you write elegant MR jobs. The steps are

1. Launch Hadoop single node/multi-node.
2. Create input.txt file.
3. Load input.txt onto HDFS.
4. Make sure Hadoop environment and Eclipse use the same java version.
5. Write MR wordcount job.
6. Add Hadoop dependencies with your project (we usually define in pom.xml) in Eclipse.
7. Create a jar file.
8. Copy jar file to the gateway node.
9. Run MR job.
10. Review job output.

3.4.1 MOVE BIG DATA INTO HDFS

1. I started Hadoop single node service.
2. Create input.txt file.

   $ vi input.txt

   ```
   hi how are you?
   ```

3. Upload input.txt onto HDFS.

   $ hadoop fs -mkdir /data
   $ hadoop fs -copyFromLocal input.txt /data
   $ hadoop fs -ls /data

3.4.2 WRITING MR JOB

I used Eclipse in Windows and moved the jar to Hadoop cluster.

4. Make sure Java version is the same in Eclipse and Hadoop environment.
5. Let us write an MR wordcount job in Eclipse and move into Hadoop cluster. Launch Eclipse in Windows (if you use Linux, it is fine) and follow the instructions given below.

   File → new → java project → project name: MapReduce → Finish

   Right click on project → new → class → name:
   WC → copy the following code [4]

```
import java.io.IOException;
import java.util.Iterator;
import java.util.StringTokenizer;
import org.apache.hadoop.conf.Configured;
import org.apache.hadoop.fs.Path;
import org.apache.hadoop.io.IntWritable;
import org.apache.hadoop.io.LongWritable;
import org.apache.hadoop.io.Text;
import org.apache.hadoop.mapred.FileInputFormat;
import org.apache.hadoop.mapred.FileOutputFormat;
import org.apache.hadoop.mapred.JobClient;
import org.apache.hadoop.mapred.JobConf;
import org.apache.hadoop.mapred.MapReduceBase;
import org.apache.hadoop.mapred.Mapper;
import org.apache.hadoop.mapred.Reducer;
import org.apache.hadoop.mapred.OutputCollector;
import org.apache.hadoop.mapred.Reporter;
import org.apache.hadoop.util.Tool;
import org.apache.hadoop.util.ToolRunner;
public class WC extends Configured implements Tool{
   public static class UDFmapper extends MapReduceBase
   implements Mapper<LongWritable, Text, Text, IntWritable>
   {
       private final static IntWritable one = new
       IntWritable(1);
       private Text word = new Text();
       public void map(LongWritable key, Text value,
       OutputCollector<Text, IntWritable> output, Reporter
       reporter) throws IOException {
           String line = value.toString();
           StringTokenizer tokenizer = new
           StringTokenizer(line);
           while (tokenizer.hasMoreTokens()) {
               word.set(tokenizer.nextToken());
               output.collect(word, one);

                   }
               }
   }
   public static class UDFreducer extends MapReduceBase
   implements Reducer<Text, IntWritable, Text, IntWritable>
   {
       public void reduce(Text key, Iterator<IntWritable>
       values, OutputCollector<Text, IntWritable> output,
       Reporter reporter) throws IOException {
           int sum = 0;
           while (values.hasNext()) {
               sum += values.next().get();
             }
```

```
                output.collect(key, new IntWritable(sum));
        }
    }
    public int run(String[] args) throws Exception {
        if(args.length <2) {
            System.err.println("Usage:<inputpath>
            <outputpath>");
            System.exit(-1);
        }
        JobConf job = new JobConf(WC.class);
        job.setJarByClass(WC.class);
        job.setMapperClass(UDFmapper.class);
        job.setReducerClass(UDFreducer.class);
        job.setOutputKeyClass(Text.class);
        job.setOutputValueClass(IntWritable.class);
        FileInputFormat.addInputPath(job, new
        Path(args[0]));
        FileOutputFormat.setOutputPath(job, new
        Path(args[1]));
        JobClient.runJob(job);
        return 0;
    }
    public static void main(String[] args) throws Exception
    {
        int exitCode = ToolRunner.run(new WC(), args);
        System.exit(exitCode);
    }
}
```

In general, a job is a program written in some computer language. An MR job is a program written in java program. It is possible to write an MR job in other programming languages (C++, Python, Ruby, etc.) as well. There are three modules in a typical MR job: map function, reduce function, and driver (main) function. In map function, pre-processing logic is coded. In reduce function, core algorithms concepts are implemented. These map/reduce functions and MR properties are invoked from driver function. In general, MR job execution begins by executing the driver function, which controls the MR execution sequence in order. Do not worry, MapReduce program constructs and art of writing MapReduce jobs are discussed in Sections 6.7 to Section 6.9.

6. Add Hadoop supporting jars to MR program from the following locations:

   Right click on the project → select properties → java build path → libraries → add external jar

   hadoop-1.2.1\lib → select all jars.

   hadoop-1.2.1\ → select hadoop-client-1.2.1.jar and hadoop-core-1.2.1.jar

7. Create JAR

   Right click on your project → export → java → JAR → select project name "MapReduce" → specify a location → finish. Give "Job" as JAR name.

8. Move Job.jar into any one of the nodes in a Hadoop cluster (use WinSCP to move jar file into Hadoop node).

9. To launch MR job from the node where Job.jar was moved, the syntax is

   ```
   $ hadoop jar jarname.jar driver_class_name /input_file_location /
   output_directory
   $ hadoop jar Job.jar WC /data/input.txt /output
   ```

   Note: Every time you run, you have to give a different output directory name (which should not exist) in HDFS. Go to WUI to see job progress, statistics, and other HDFS information.

   NN_IP:50070 to track HDFS details
   JT_IP:50030 to track JT details

10. The output is created in HDFS along with two other files in HDFS output folder

    ```
    $ hdfs dfs -ls /output
    ```
`_SUCCESS`	– It is an empty file to indicate job success
`_logs`	– It is a directory that contains job history
`part-r-00000`	– It is the output file

    ```
    $ hadoop fs -cat /output/part-r-00000   // to display output
    ```

You can load data onto HDFS and launch a job from any node in the cluster as job client runs in every Hadoop node. Because we have set mapred.xml and core-site.xml in all nodes. Therefore, the job request is handed over to JT. Try running the MR program with a small amount of data first in local mode setup, so it will be easy for you to detect any errors and verify the logic of your program. Then go for launching in a multi-node cluster with huge data.

3.5   HDFS AND MR PROPERTIES

Some common properties that are mostly used to customize MR execution sequence are discussed below:

core-site.xml

fs.default.name → specifies the NN.

hadoop.tmp.dir → specifies a location to store FSImage and edit logs in NN, and blocks in DNs. It contains comma separated paths to set multiple locations/disks.

webinterface.private.actions → adds a button that help users to kill running jobs from WUI. In a production environment, users should not be allowed to kill running jobs accidentally.

hdfs-site.xml

dfs.replication → indicates RF of a block. It can be set for each file. RF must always be less than or equal to the number of DNs available in the cluster.

dfs.block.size → specifies the block size in bytes. Default block size is 67108864 bytes (64 MB).

dfs.name.dir → specifies the location in NN to store FSImage.

dfs.data.dir → specifies the location in DN to store HDFS blocks.

fs.checkpoint.dir → specifies the location in SNN to perform checkpointing.

dfs.namenode.handler.count → specifies the number of threads in NN to serve requests, default is 10.

dfs.datanode.handler.count → specifies the number of threads in NN to serve requests, default is 3.

dfs.datanode.failed.volumes.tolerated → specifies the number of local disks of a DN die before failing the entire DN. Default is 0.

dfs.hosts.exclude → indicates a list of DNs IP to remove from the cluster.

mapred-site.xml

mapred.job.tracker → specifies the JT IP address.

mapred.child.java.opts → used to specify JVM heap size in memory for a map and reduce task. -XmxNu denotes the maximum JVM heap size in memory and -XmsNu denotes the initial JVM heap size. Default is -Xmx200m, but recommended is -Xmx1g. By default, every TT is assigned with two map slots and two reduce slots. Therefore, we need at least (2+2) * 200 = 800 MB or more memory in each TT.

mapred.output.compress → for best optimal performance, we should mini-mize the network traffic by using compression. Set 'true' to this property.

mapred.output.compression.codecorg.apache.hadoop.io.compress.Snappy-Codec (default).

mapred.output.compression.type → compression is done either record-level or block-level. RECORD is the default option, but BLOCK is usually recommended. NONE means no compression.

mapred.local.dir → specifies the path to store intermediate results of map tasks.

mapred.system.dir → specifies the path to store shared files for all map/reducer tasks.

mapred.map.tasks → denotes the number of map tasks that can be run. By default, the number of map tasks is controlled by the block size and IS size.

mapred.reduce.tasks denotes the number of reduce tasks. Default is 1, which is very inefficient and hence will not take any advantage of the parallel execution of Hadoop.

mapred.tasktracker.map.tasks.maximum → 2 (number of map task slots per TT by default).

mapred.tasktracker.reduce.tasks.maximum → 2 (number of reduce task slots per TT).

mapred.reduce.parallel.copies → number of map task's output a reducer can read in parallel. Default is 5.

mapred.reduce.slowstart.completed.maps → specify after how much % of map task completion, reduce tasks can be launched. Default is 0.05 (5%). Value of 1 will instruct reduce tasks to wait for all the map tasks to finish before starting. Value of 0.0 will start all reduce tasks along with mappers right away. A value of 0.5 will start the reduce tasks when half the number of the map tasks are completed.

io.sort.mb → buffer size to store intermediate result in memory. Default is 100 MB.

io.sort.spill.percent → Default value is 0.8. When the in-memory buffer is filled up to 80%, a thread begins, partitions, sorts, and writes the intermediate result to local disk.

io.sort.record.percent → percentage of io.sort.mb dedicated to tracking record boundaries. The default value is 0.05.

io.sort.factor → defines the number of partitions to merge for sorting. Default is 10.

mapred.jobtracker.taskScheduler → default is org.apache.hadoop.mapred. FairScheduler.

3.6 HADOOP ADMINISTRATIVE COMMANDS

$ hadoop namenode -format	Formats HDFS and creates namespace.
$ hadoop secondarynamenode	Interacts with SNN.
$ hadoop namenode	Interacts with NN.
$ hadoop datanode	Interacts with DN.
$ hadoop dfsadmin	Interacts with NN admin client.
$ hadoop mradmin	Interacts with MR admin client.
$ hadoop fsck	Checks distributed file system.
$ hadoop fs	File system shell to interact with HDFS.
$ hadoop balancer	Balances the DN's load.
$ hadoop jobtracker	Interacts with JT.
$ hadoop pipes	Runs a pipes job.
$ hadoop tasktracker	Interacts with TT.
$ hadoop historyserver	Interacts with job history servers.
$ hadoop job	Interacts with MR jobs.
$ hadoop queue	Job queue information is displayed.
$ hadoop version	Prints the Hadoop version.
$ hadoop jar	Executes MR job jar file.
$ hadoop distcp <srcurl> <desturl>	Copies file or directories across cluster.
$ hadoop classpath	Displays the path containing jar file and required libraries.
$ hadoop daemonlog	Get/set the log-level for each daemon.

3.7 COMMANDS TO INTERACT WITH MR JOBS

$ hadoop job -list all	Displays all incomplete jobs meta-data (job ID).
$ hadoop job -submit <job-file>	Submits the MR job.
$ hadoop job -status <job-id>	Displays the job completion status and counters.
$ hadoop job -counter < job-id> <group-name> <counter-name> Displays the counter value.	

$ hadoop job -kill <job-id>   MR job is terminated.
$ hadoop job -history [all]<jobOutputDir> -history <jobOutputDir>
                                     MR job details are displayed from job
                                     history.
$ hadoop job -kill-task <task-id>   To terminate map/reduce task.
$ hadoop job -set-priority <job-id> <priority> Job priority can be set.
                                     Priority values are VERY_
                                     HIGH, HIGH, NORMAL,
                                     LOW, VERY_LOW

KEYWORDS

- **Hadoop 1.2.1 single node and multi-node installation**
- **HDFS and MapReduce commands**
- **writing MapReduce jobs**

CHAPTER 4

Hadoop Ecosystem

Creativity is intelligence having fun.
—Albert Einstein

INTRODUCTION

Hadoop is mainly known for two core tools: HDFS and MR. Today, in addition to HDFS and MR, the term represents several sub-projects. Figure 4.1 gives you an overview of Hadoop sub-projects for various purposes. Let us have a glance at some of the Hadoop sub-projects for big data processing. Go to *incubator.apache.org* and *incubator.apache.org/projects* to explore further.

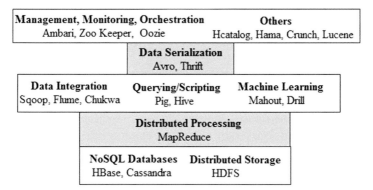

FIGURE 4.1   Hadoop sub-projects.

Before we start processing big data, it has to be loaded onto HDFS or HBase using data ingestion tools like Sqoop, Flume, etc. Data sometimes is streamed from sources like sensor devices, web pages, disks, etc. using tools like Kafka to the data processing tools (Figure 4.2) such as MapReduce,

Spark, etc. These data processing tools help us in writing logic to extract information, which can then be displayed to the audiences in a presentable manner using visualization tools such as Kibana, Tableu, and the like. Nowadays, MapReduce is used behind Matlab for distributed data processing. Matlab users install an interface that takes care of transforming Matlab codes into a series of MapReduce jobs.

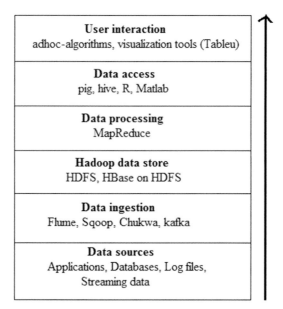

FIGURE 4.2   Big data processing sequence.

4.1 DATA PROCESSING MODEL

There are different distributed data processing models, as shown in Figure 4.3 [3]. Some of these tools overlap each other at a minimum-level.

1. MapReduce model

We discussed enough of MapReduce model. It is a data-parallel batch processing tool that comprises a set of map and reduce tasks. Aggregation-related applications such as counting, ranking are typical use cases of MapReduce. Example: Google MapReduce (2004), Hadoop MapReduce (2005), Apache spark (2009).

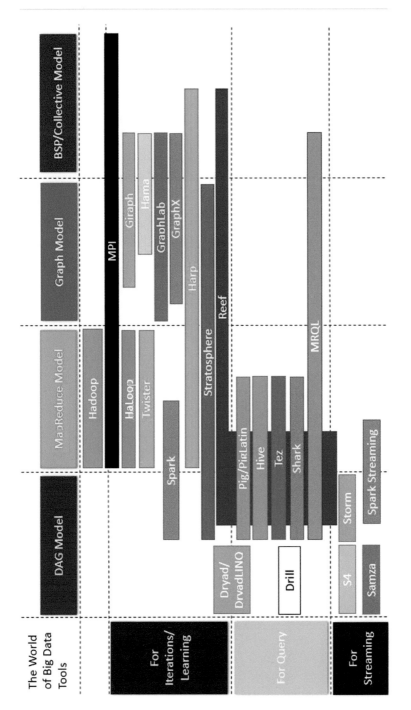

FIGURE 4.3 Different data processing model for big data. (Reprinted from Ref. [3] with permission).

2. Graph processing model

A graph contains a set of vertices and edges. Graph data is divided and loaded into multiple nodes and processed in parallel. It involves significant computational dependencies (not as much as BSP) and multiple iterations to converge. It does not require any global synchronization. Graph operations using MapReduce typically need one job per iteration. It requires the entire graph data structure to be serialized for each iteration. If there are 100 iterations to process the entire graph, 100 MapReduce jobs are executed, which results in huge latency. Therefore, there are some specialized tools to work with colossal graph data. Example: Pregel (2010), Apache Hama (2010), Apache Giraph (2012). Figure 4.4 shows the difference between the general graph and directed acyclic graph.

3. Directed Acyclic Graph (DAG) model

The DAG contains a finite number of vertices and directed edges that form no cycle. Every edge in the graph moves forward. For graph-based applications, we will have to write more than one MapReduce job and coordinate their executions. So, the MapReduce model is not suitable for graph-based applications as it involves large IO. However, some tools provide support for graph-based applications and give an interactive, real-time response. Example, DryadLINQ (2007), Pig/Hive, Spark, Tez, etc.

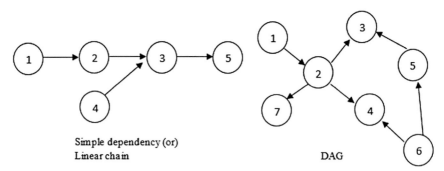

Simple dependency (or)
Linear chain

DAG

FIGURE 4.4 Linear graph vs DAG.

4. Collective/Bulk Synchronous Processing (BSP) model

Large dependencies among vertices in a graph require communication, coordination, and global synchronization. It is called BSP, which is typical in scientific computations. Some of the tools for BSP are Harp (2013), Google Pregel, Apache Giraph, and Hama.

4.2   QUERYING AND SCRIPTING

Data must be structured or organized in order to perform query processing. Tools that support querying big data are Google Sawzall (2003), Apache Pig (2006), Apache Hive (2007), Dremel (2010), Shark (2012), Apache Drill (2012), Apache Tez (2013), Impala.

1. Pig

It is a friendly tool for non-java programmers to process big data and developed by Yahoo. 50% of jobs in Yahoo are implemented using Pig. 100 lines of MapReduce program in Java can be written in just 10 lines using Pig. However, you do not have much flexibility for writing user-defined algorithms as there are a limited number of predefined queries. Pig stores and manages huge structured/semi-structured data in HDFS. It uses Pig Latin language to process data. Pig runtime compiles and converts pig queries to low-level MapReduce jobs at runtime.

2. Hive

It is mainly for SQL users who are non-java programers and provides DWH infrastructure on top of HDFS. Relational DWH is centralized, but the Hive is a distributed DWH on top of HDFS. It uses Hive Query Language (HQL), very similar to SQL, to work with structured data. Similar to Relational DWH, you have to define the schema and then load huge data onto Hive. Hive runtime converts HQL to a sequence of low-level MapReduce jobs. Both pig and hive are not client-server architecture and mainly used to perform ETL based pre-processing before applying any algorithms.

3. Impala

Hive provides distributed DWH facility on top of HDFS. Hive does not guarantee a real-time response. To overcome this, Impala was introduced by Cloudera. It is a SQL-based tool for a real-time response on HDFS. It is because, typical hive query is transformed into a series of MapReduce jobs that are executed in sequence and combines the result, which takes more time. Impala bypasses MapReduce altogether and performs distributed query processing by running its daemon on each slave node in the cluster.

4. Apache Tez

Tez supports key-value data model of MapReduce and tuple-based data model of Hive. It is used to build complex DAG of tasks with Pig and Hive.

It also supports reduce operation without sorting overhead that MapReduce requires.

4.3 DATA STORAGE (NOSQL)

NoSQL database systems were developed to provide operational functionalities (insert, delete, update) with the interactive and real-time response from big data, unlike HDFS. These NoSQL databases support one of the four data models (key-value, column-oriented, document, graph).

1. HBase

It is a column-oriented database based on Google Big Table (huge structured database built on top of GFS). Unlike pig and hive, it provides an interactive response on HDFS. It supports tables with millions of columns and billions of rows on top of HDFS. Data is indexed before performing analytics. Only HBase in all NoSQL database systems is closely integrated with MapReduce and HDFS. Pig and Hive also natively work with HBase. HBase works on master-slave architecture and supports horizontal scalability, consistency, and auto-failover. MapReduce jobs can be run on HBase data.

2. Cassandra

It is a real-time, interactive, column-oriented database based on Dynamo and HBase databases. Cassandra works on multi-master architecture and suitable for accessing data across multiple clusters.

3. Hana

It is an in-memory, column-oriented database for processing huge data in real-time. Hana is ACID properties compliant, supporting SQL.

4.4 DATA SERIALIZATION

Data serialization is a process of packing structured/unstructured objects into a common format and writing it on stream to either store or exchange them across heterogeneous applications in a compact way. Serialized data formats are typically in JSON, XML, BSON data formats. Typical data serialization tools are Avro and Thrift. Sometimes map task is written in Java and reduce task is written in Python. Java and Python use different data types that are not compatible with each other. So, we use Avro or Thrift to achieve this cross-language compatibility.

1. Avro

Apache Avro is a language-neutral data serialization system. It models the data that can be processed in many languages (C, C+, C#, Java, Python, Ruby, etc.). Avro assumes that the schema (written in JSON) is always present during read/write time.

4.5 DATA INTELLIGENCE

Typical machine learning algorithms require multiple iterations to reach a final result. Developing scalable machine learning algorithms is very difficult, and there exists a library containing already implemented algorithms. Example, Mahout on MapReduce, MLBase on Spark, Ricardo, presto.

4.6 STREAM PROCESSING

Stream processing tools process unbounded data and give a response in real-time. They process data as it arrives at the computer and provides fresh low latency results before the data persistently stored. Example: S4 (2011), Storm (2011), Spark streaming (2012), Samza (2013), Kinesis, Kafka. Spark dominates Storm because it is a single programming paradigm for both offline (batch) and real-time data analytics. Once you master Spark, you can work on a variety of tools.

4.7 DATA INTEGRATION TOOLS

The HDFS commands are used to load data from a local file system onto HDFS. There are tools to load data from other sources such as RDBMS and streams.

1. Sqoop (SQL+ Hadoop)

This integrates Hadoop with relational database systems. You cannot perform massive data processing in relational databases. Therefore, huge structured data is moved into HDFS for analysis. Sqoop also converts its queries to a series of MapReduce jobs and executes at a low-level. Sqoop performs EL in ETL.

2. Chukwa and Flume

Log processing is one of the suitable applications of MapReduce. Logs are generated monotonically across many machines and moved dynamically onto HDFS in real-time using tools like Chukwa, Flume.

4.8 MANAGEMENT, MONITORING, ORCHESTRATION

Every big data processing tool comprises a set of services to be deployed across the cluster. Similarly, every application comprises a set of jobs to be executed differently. These two paradigms require some assistance like synchronization and coordination to improve performance. There are some tools to achieve this at the tool-level and application (job)-level.

1. Zookeeper

It is very similar to Chubby in Google and performs distributed service coordination and synchronization. It is mainly used in HBase, NN HA.

2. Oozie

Manual effort is required to run and coordinate a set of MapReduce jobs for graph-based applications. Oozie is a workflow scheduling library to manage MapReduce jobs in DAG. It helps in scheduling jobs one after another in a sequence. Jobs are triggered by time and data availability. Falcon is also a workflow scheduling library.

3. Ganglia and Nagios

Ganglia is used to collect cluster metrics, and Nagios is used for system alerts. Nagios will alert you by sending emails/messages when your attention is needed in a situation like a node crash, disk space low, etc.

4. Ambari

It is a remote deployment and management tool with nice WUI. Ambari provides a step-by-step wizard for installing Hadoop services across the cluster and central management for starting, stopping services. It displays a dashboard for monitoring the health and status of the Hadoop cluster.

5. Hue

It is a web-based, Hadoop tools management service. It provides a single portal for different Hadoop tools and eases user interaction with those tools.

Therefore, you need not use CLI or other complex WUI for every Hadoop tool. For example, it is easy to upload data onto HDFS using Hue than using the command line.

4.9 OTHER DISTRIBUTED BIG DATA PROCESSING TOOLS

Nutch is an Opensource web search engine from Yahoo.

Lucene is a text search library.

Solr is a high-performance search engine built on Lucene. It accumulates huge documents on HDFS with index facility to search document contents quickly.

UIMA is an unstructured information management system architecture for development, discovery, composition, deployment, and analysis of unstructured data.

Knox provides centralized security in Hadoop.

HDT is Hadoop development tools to integrate Hadoop with Eclipse. It helps you to launch MR jobs into Hadoop cluster from Eclipse.

Apache Wink provides a framework for building RESTful web services.

Apache Whirr provides a set of libraries for running cloud services.

Sahara or Savanna is a project to integrate Hadoop with OpenStack.

HCatalog exposes hive meta-data to other Hadoop applications.

Apache Crunch is a library that helps in writing and testing user-defined MapReduce pipeline.

Apache Slider makes it possible to run a different version of the same application and provides control to dynamically change the number of nodes an application is running.

Apache Twill is very similar to slider but has a programming model for developing applications to run on YARN. This also provides real-time logging and messaging facility.

Data visualization tools such as D3, Pentaho, Tableau, etc. help to visualize huge data loaded onto HDFS. Huge data cannot be viewed in excel as it takes more time to load in memory.

4.10 DECISION SUPPORT SYSTEM (DSS) FORECAST

Faster data analytics directly affects the bottom line of business and managerial decision making at all levels of an enterprise. Therefore, the DSS such as Hadoop for unstructured data and DWH for structured data grow rapidly.

Hadoop MapReduce market forecast 2015–20 [5]

Hadoop MapReduce market is set to grow at a compound annual growth rate of 58.2%, reaching $50.2 billion in 2020. It would not be an exaggeration to say that today, Hadoop MapReduce is the only cost-sensible and scalable big data management.

DWH acceleration market forecast 2015–20 [6]

It is predicted to grow at a compound annual growth rate of 23.3%, reaching $3.85 billion by 2020. The DWH acceleration technologies such as Field Programmable Gate Array (FPGA), General Purpose computing on GPU (GPGPU) are currently emerging and much of the necessary knowledge, infrastructure and experience are already in place, though not widely.

4.11   HIGH-PERFORMANCE DATA ANALYTICS (HPDA)

A large number of servers in a cluster or a large number of cores in a processor (GPU) that perform over TFLOPS is called a typical HPC. It is mainly used for compute-intensive applications. Popular implementations to achieve HPC are MPI (1992), MPICH (2001), OpenMP (2004), charm++, Linda, PVM, condor. The HPC is offered as a service by different cloud service providers: Amazon Web Services (AWS), Google cloud platform, Microsoft Azure big compute, IBM spectrum computing, Penguin computing on demand, etc.

The HPC requires to processes huge data for today's advanced applications like deep learning. This is called HPDA. Some active fields in HPDA are listed below.

1. Genomics is the study of genes (the genome), and its interactions with other genes and environment. Genomics involves two core activities:
    - Sequencing: reading DNA from the cells of an organism and digitizing it.
    - Computation: gene sequence alignment, compression, etc.
2. Oil and gas exploration also require compute-intensive simulations.
3. Ray Tracing is widely used in image processing applications.

4.12   HADOOP VS SPARK

Spark is built to overcome the certain limitations of MapReduce. Hadoop was developed by Apache and Spark was initially developed at UC Berkeley later it was donated to Apache. MapReduce and Spark are entirely different programming paradigms. MapReduce requires data to be persistently stored in disks before processing. However, spark processes keeping data in memory before persistently stored in disks, which guarantees the real-time response for stream processing. This is the significant difference at the architecture-level in these tools. Moreover, Spark has special functions along with the classical map and reduce functions that are in MapReduce.

Many reports claim that Spark is 10 times faster than MapReduce at the disk-level and 100 times faster than MapReduce at the memory-level. To exploit this performance, one should be ready to invest more for high performing RAM. The ultimate objective of MapReduce is to transform a cluster of commodity servers and our desktops/laptops into big data processing cluster. However, commodity servers are not enough for Spark to obtain maximum performance. Therefore, MapReduce is cheaper and better for low-level organizations and people who cannot afford costly RAM. Also, Spark is expensive to consume as a service from the cloud. In short,

- Spark is a better option if we work on real-time applications and machine learning algorithms, but a little bit costlier.
- MapReduce is a better option for applications which is not time-bound.

To conclude, MapReduce will co-exist with Spark for a long period because Spark does not have any storage and resource management layer. Spark uses HDFS and YARN tools from Hadoop. So, Spark needs to come a long way to replace Hadoop. It may be safe to say that Spark could be integrated with Hadoop, rather than have it as a replacement.

4.13   HADOOP V3

Apache Hadoop has undergone significant changes to adapt more features and fix bugs over Hadoop v2. We will see the novel features that make Hadoop v3 better than its predecessors Hadoop v2 and Hadoop v1.

1. Minimum runtime version for Hadoop v3 is JDK 8

As Java 7 ended its life in 2015, there was a need to revise the minimum runtime version to Java 8 with a new Hadoop release so that the new release is supported by Oracle with security fixes and also will allow Hadoop to upgrade its dependencies to modern versions.

2. Support for Erasure Coding in HDFS

Replication (by default 3) provided by HDFS is costly in Hadoop v2, and it requires 200% additional storage space. Hadoop v3 incorporates Erasure Coding in place of replication consuming comparatively less storage space while providing the same level of fault tolerance. Erasure Coding is a 50 years old technique that lets any random piece of data to be recovered based on other piece of data. With support for erasure coding in Hadoop v3, the physical disk usage will be cut by half (3x to 1.5x), and the fault tolerance will increase by 50%.

3. Hadoop shell script rewrite

With Hadoop v2 shell scripts were difficult to understand what is the correct environment variable to set an option and how to set (in java.library.path or java classpath or GC). To address several long-standing bugs and provide unifying behaviors and enhance the documentation and functionality, Hadoop shell scripts are rewritten in Hadoop v3.

4. MapReduce task-level native optimization

A new native implementation of the map output collector to perform the sort, spill, and serialization will improve the performance of shuffle-intensive jobs by 30%.

5. Support for multiple NNs to maximize fault tolerance

With support for only two NNs, Hadoop v2 did not provide the maximum level of fault tolerance, but in Hadoop v3, there will be additional fault tolerance as it offers multiple NNs. This new feature is just perfect for business-critical deployments the need to run with high fault tolerance levels.

6. Powerful YARN in Hadoop v3

The YARN was introduced in Hadoop v2 to make Hadoop clusters run efficiently. In Hadoop v3, YARN has come off with multiple enhancements:

Better resource isolation for disk and network, resource utilization, user experiences, and elasticity.

7. Change in default ports for various services

The default port numbers for NN, DN, SNN have been moved out of the Linux ephemeral port number range (32768-61000) to avoid bind errors on startup because of conflict with other application.

KEYWORDS

- **big data processing models**
- **Hadoop ecosystem**

CHAPTER 5

Hadoop 2.7.0

Do what you can with all you have, wherever you are.

—Theodore Roosevelt

INTRODUCTION

It is recommended to read Chapter 2 to refresh some basic terminologies and concepts before beginning Hadoop 2.x. If MRv1 is installed in a cluster, no other data processing framework can be deployed. Because, it does not share the cluster resources to other co-located data processing framework such as stream processing, in-memory computing, etc. So, Hadoop 1.x framework is good enough only for MapReduce (MR). Therefore, Hadoop 2.x splits MRv1 functionalities into two software components (MRv1 and YARN) to exploit more scalability and resource sharing among frameworks, as shown in Figure 5.1.

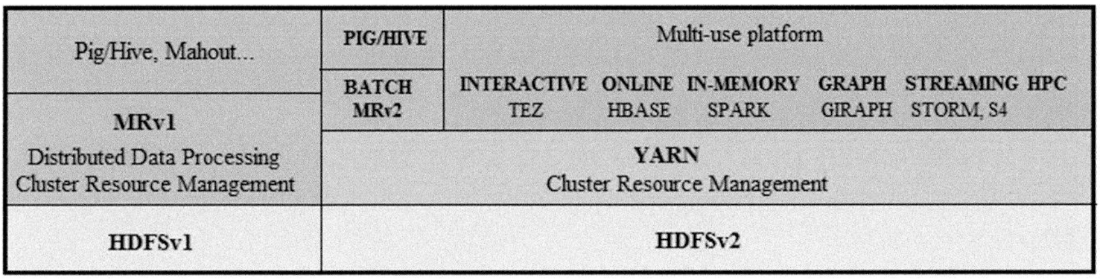

Pig/Hive, Mahout...	PIG/HIVE	Multi-use platform					
	BATCH MRv2	INTERACTIVE ONLINE IN-MEMORY GRAPH STREAMING HPC					
MRv1		TEZ	HBASE	SPARK	GIRAPH	STORM, S4	
Distributed Data Processing Cluster Resource Management	YARN Cluster Resource Management						
HDFSv1	HDFSv2						

FIGURE 5.1   Hadoop 1.x vs Hadoop 2.x.

The YARN is installed in the cluster of servers to share resources among many tools and frameworks. To achieve this, all tools are reworked to run on YARN as YARN applications. MRv1 also was reworked at the architectural-level to run on YARN as YARN application and renamed to MRv2. The HDFS is the only common tool used in the cluster for both Hadoop v1 and v2. Some significant Hadoop 2.x features are:

- Sharing cluster resources among tools/frameworks.
- High scalability.
- HA and load balancing for NN.
- Improved cluster utilization.

5.1 HDFS V2 FEATURES

The HDFSv2 has many advantages over HDFSv1: NN HA to avoid SPOF, HDFS federation to load balance NN, and improved data security.

HDFS federation

In HDFSv1, the NN maintains the whole meta-data in memory. If meta-data grows out of memory, we have to increase the memory size, else the number of blocks that can be managed in HDFS will be less. To overcome this limitation, Hadoop 2.x has introduced HDFS federation, which is a group (federation) of multiple independent NNs to balance the load of a NN.

NN HA

The NN HA also includes more than one NN to overcome SPOF. In HDFSv1, to recover NN from failure, it takes over 30 minutes in a production environment as it involves manual intervention. The NN HA provides auto-failover which avoids manual intervention to bring up new NN functioning. This feature eliminates the downtime (only a few minutes) and makes the HDFS cluster remain functional.

5.2 COMPONENTS IN HADOOP 1.X VS HADOOP 2.X

MRv1

JT is responsible for (as shown in Figure 5.2):

- resource management (tracking resource consumption/availability in TTs).
- job life-cycle management (scheduling jobs, tracking the progress of jobs/tasks).
- providing fault-tolerance for tasks.
- job history management.

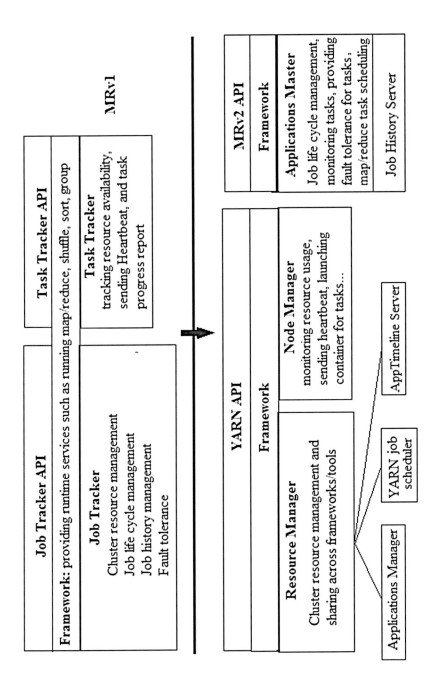

FIGURE 5.2 MRv1 vs MRv2 components.

TT has simple responsibilities:

- launch map/reduce tasks on order of JT.
- send resource availability and task progress to JT.

API is a library that contains a set of pre-coded functions which can be used ready-made to interact with components like resource manager, node manager, etc.

MR framework takes care of launching map/reduce tasks, running shuffle + sort + group.

MRv2

The primary short-comings with MRv1 are:

- JT is not able to support beyond a certain number of task execution simultaneously
- JT is unable to share resources of a cluster with co-located non-MR data processing frameworks due to fixed slot size.

Therefore, the functionalities of JT in MRv1 is split into multiple components, as shown in Figure 5.2 and Table 5.1.

TABLE 5.1   MRv1 vs MRv2 vs YARN Components

MRv1	MRv2	YARN		
Job Tracker (JT)	MR Application Master	Resource Manger (RM)		
	Job History Server	Applications manager	YARN application (job) scheduler	AppTimeline Server
Task Tracker (TT)	-	Node Manager (NM)		
Slot	-	Container		

MRv2 advantages

- It is backward compatible with existing MRv1 and allows MRv1 tasks to execute.
- It results in better cluster utilization by providing capacity guarantee and fairness among jobs of different tools.

YARN advantages

- It allows deploying any tools and frameworks (multi-tenancy) that are compatible with YARN.

- Every framework tends to do the job scheduling, log aggregation, etc. However, YARN performs job scheduling, log aggregation, etc. on behalf of YARN applications.

5.3 HADOOP 2.X COMPONENTS IN DETAIL

Figure 5.3 illustrates the major components of YARN and MRv2 in Hadoop physical cluster mode. These components are run as a service in one or more server and discussed in detail in the following sections.

1. Yet Another Resource Negotiator (YARN)

The YARN pools cluster resources and shares them across tools and frameworks. YARN is essentially a software system for managing different distributed frameworks. It manages and shares cluster resources in a fine-grained manner (sharing resources in any proportion upon its availability) among different frameworks deployed in the same cluster for better cluster utilization. The YARN does not know what kind of application is running on it. It can be either the job of MR or Spark or Storm. The YARN is based on master-slave architecture and has two primary components: resource manager and node manager.

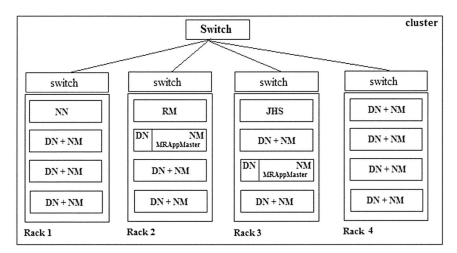

FIGURE 5.3   MRv2 components in Hadoop cluster.

a. Resource Manager (RM)

It is a master service and a centralized resource manager in the cluster. It instructs node manager to launch containers for map/reduce tasks. There is only one RM in the cluster. Specific RM functionalities are:

- managing and sharing cluster resources.
- handling resource requests.
- providing security with Kerberos.

RM subcomponents are:

- Applications Manager
- YARN Scheduler
- AppTimeLine Server

Applications manager (AsM)

It is responsible for accepting the job-submissions, negotiating a container for executing MRAppMaster. It provides fault-tolerance for MRAppMaster.

YARN scheduler

In general, every processing framework has its job scheduler. When these frameworks are installed on YARN, they need not do scheduling anymore. As YARN manages resources in the cluster, it effectively and efficiently schedules the jobs of different tools and frameworks. The YARN scheduler is responsible for allocating resources to jobs of different frameworks subject to the capacity, queue, etc. MRv2 is just a programming paradigm and does not do job scheduling. Since different frameworks are hosted on YARN, jobs of MR/Spark/Storm are called as YARN applications with the global application ID.

AppTimeline server

AppTimeline server is a global job history management for all jobs from different frameworks running in the cluster. So, frameworks need not do job history management.

b. Node Manager (NM)

Every slave node runs an NM daemon, which manages slave resources (memory and CPU). The NM communicates with RM to register itself to be a part of a YARN cluster. Subsequently, NM periodically sends resource availability to RM via HB and responds to the instructions from RM and MRAppMaster. The NMs responsibilities are creating, monitoring, and killing containers for MRAppMaster, map, and reduce tasks on the instructions from RM. It monitors the resource usage of containers and kills them if they try to consume more resources. The NM also provides local logging services to applications. The NM runs auxiliary services such as shuffle, sort, and group during task execution for MR jobs. The NM launches two general services along with this.

YARN client – is a lightweight process that interacts with AsM to submit a job and to store job meta-data in HDFS.

YARN child – is a lightweight process running in containers. Its particular responsibilities are to localize task-related information from HDFS and send task status information to MRAppMaster periodically. In contrast, in MRv1, TT sends task status report to JT.

c. Container

The container is a basic unit of resource allocation in YARN for any applications/jobs. A container is composed (see Figure 2.13) of virtual CPU and a portion of memory to launch map/reduce tasks and MRAppMaster. A container is scheduled by RM and supervised by NM. In MRv1, users can define the number of map/reduce slots and its configuration in TTs. However, the serious disadvantage of a slot-based scheme is, MR services should be stopped to reconfigure the slots. Moreover, the slots are not job-specific. For instance, consider a map slot configured with 1 GB memory. Even if a map task requires only 200 MB, it occupies the entire 1 GB and will not release the unused memory until the task completion or termination. Therefore, it causes resource under-utilization.

In contrast, the container in MRv2 is job-specific. Every job can be assigned with different size of the container, as shown in Figure 5.4. For example, container1 (2 GB memory, 1 logical core), container2 (3 GB memory, and 2 logical cores) can have a different configuration. However, the number of containers possible in an NM depends upon the number of cores and the size of memory dedicated to YARN in the particular NM.

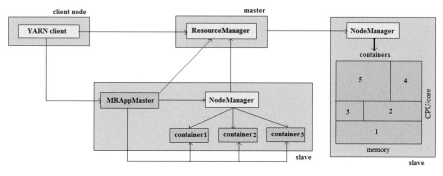

FIGURE 5.4   Container allocation.

Note: RM, NM, NN, SNN, and DN services are not launched in containers, but they run as a separate java process in the JVM.

2. MR Application Master (MRAppMaster)

The YARN provides an Application Master for each job of MR, Spark, Storm, etc. For MRv2, there is an MRAppMaster managed by AsM. It is created when an MR job begins and destroyed when an MR job ends. Every MR job has its own dedicated MRAppMaster. If two MR jobs are running in the cluster, then two MRAppMaster will be running (see Figure 5.3). MRAppMaster manages MR job life cycle (in MRv1 JT does job life cycle management). It requests RM for containers to launch map/reduce tasks via HB message. RM response to the HB contains information on the containers to be launched. If a task requires more resources, YARN child of that task requests MRAppMaster, which in turn requests RM. The YARN child periodically sends task status information (progress/success/failure) to MRApp-Master. MRAppMaster coordinates map/reduce tasks and phase, aggregates logs and counters from NMs. It behaves like a short-living JT for each MR job. It provides fault-tolerance for map/reduce tasks. MRAppMaster itself handles most of the MRv1 functionality.

3. Job History Server (JHS)

It is a daemon that serves historical information about completed MR jobs. It archives job's metrics and meta-data. Every NM records task activities locally. Whenever an MR job is submitted, JHS will give you an MRAppMaster URL (in the console while starting the job) to track the running application. Because YARN client does not know in which node MRAppMaster is running. Once the job is done, log information of that particular job from various NMs is moved to JHS. MRv1 logs are scattered across nodes and not accumulated in a centralized place. However, in MRv2, we maintain a central JHS to manage MR job logs. JHS is only for MR jobs. However, YARN uses AppTimeline server to maintain logs of all the applications regardless of frameworks.

4. YARN Job Scheduler

In MRv1, job and task scheduling are done by MR scheduler itself. In MRv2, job scheduling is done by YARN scheduler, and MRAppMaster does task scheduling. The YARN scheduler schedules and allocates resources for not only MRv2 jobs and also for the jobs of all other YARN applications. It is also pluggable as in MRv1. The YARN scheduler does not monitor task progress and does not provide fault-tolerance for tasks because it is not application specific. So, MRAppMaster is designed to track task progress and provide fault-tolerance for MR tasks.

MRv1 and YARN scheduler are distributed with the choice of FIFO, capacity, and fair schedulers. Default scheduler of MRv1 is FIFO and for YARN is the capacity scheduler. Detailed differences between Hadoop v1 and Hadoop v2 are accounted in Table 5.2.

5.4 MRV2 JOB EXECUTION FLOW

Every MR job in YARN is executed as a separate YARN application and assigned with a dedicated MRAppMaster. The MRv2 job execution flow is similar to MRv1, but the responsibilities are split and assigned to different components to achieve scalability and resource sharing among multiple frameworks.

TABLE 5.2 Hadoop v1 vs Hadoop v2

Hadoop 1.x	Hadoop 2.x
Only the MRv1 framework can be deployed in the cluster (monolithic system).	YARN shares cluster resources with other non-MR frameworks such as Strom, Spark.
Basic resource unit is a slot. Every TT contains a fixed number of map and reduce task slots. Map tasks cannot use reduce task slots and vice versa. A number of slots and its capacity are fixed. This static configuration causes resource under-utilization. If you want to change the number of map/reduce slots and its configurations, you have to stop, re-configure, and restart the cluster.	Basic resource unit is a container. A container can execute map or reduce tasks or any other jobs. A number of containers in a node are not fixed, and each job can define any size of containers. It improves resource utilization.

TABLE 5.2   *(Continued)*

Hadoop 1.x	Hadoop 2.x
MRv1 JT does both resource management, job life cycle management, and fault-tolerance.	YARN does cluster resource management, and MRAppMaster manages job life cycle and fault-tolerance.
JT does the job and task scheduling.	Job/application scheduling is done by YARN scheduler, and map/reduce task scheduling is done by MRAppMaster.
Scaling is limited to 4000 nodes per cluster and 40,000 tasks.	Scalable up to 10,000 nodes per cluster and 100,000 tasks.
Single NN manages the entire namespace.	Multiple NNs balance the load – NN federation
NN suffers from SPOF and needs manual intervention to overcome.	NN HA overcomes SPOF with auto fail-over (no manual intervention).
HDFSv1 default block size is 64 MB.	HDFSv2 default block size is 128 MB.
In MRv1, log management is not centralized.	In YARN, there is a application timeline server to maintain logs of all applications and a MR JHS to maintain history of MR jobs separately.

Figure 5.5 illustrates the execution flow of an MRv2 job in YARN. Major phases are job submission, job initialization, task assignment, task execution, status updates, and job completion. Each step given in the figure is explained below.

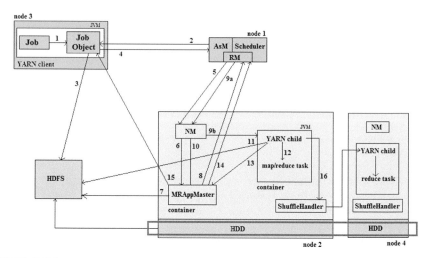

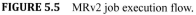
FIGURE 5.5   MRv2 job execution flow.

Job submission

1. A client submits an MR job by interacting with the Job object (Job Submitter).
2. Job object interacts with AsM and acquires application meta-data such as application ID. Job object verifies whether input files and output directory already exist by interacting NN. If an output directory exists or input files not available, IOException is thrown.

   Note: Application ID is global and assigned by YARN. The application ID is framed from the starting time of RM and incremented for upcoming jobs. MR job ID is derived from an application ID. Before MR job starts, the Job ID is printed on the console. Example:

Application ID	: application_1410450250506_0001
Job ID	: job_1410450250506_0001
Map task ID	: task_1410450250506_0001_m_000000
Map task attempt ID	: attempt_1410450250506_00 01_m_000000_0

3. Job object calculates the number of IS (enable yarn.app.mapreduce. am.compute-splits-in property) by interacting NN and moves all the job-related resources (jars, job configurations, and IS information) onto HDFS to make them available for all NMs. These information is replicated (mapreduce.client.submit.file.replication) to 10 DNs across the cluster by default.
4. Job object submits the application to AsM. The output directory is created right after job submission is successful, but output files are created only when reduce task writes its output.

Job initialization

5. Right after job submission, AsM interacts with RM to launch a container for MRAppMaster. RM directs scheduler to create a container for MRAppMaster in an NM. MRAppMaster registers itself with the RM on its start-up.
6. Calculating IS in step 3 is slower. So, you can set MRAppMaster to calculate IS. MRAppMaster decides how to run map/reduce tasks of a job. If the job is small, MRAppMaster may choose to run tasks in the same JVM itself (such a job is called uber job).
7. MRAppMaster grabs required job-related information from HDFS, such as IS information (if already calculated by job client), job configurations and the like.

8. MRAppMaster retrieves block locations from the NN to form IS. MRAppMaster assigns a task ID for both map and reduce tasks. Then, it sends resource request to RM to launch containers for map/reduce tasks. The resource-request carries the following information:

    - List of three (RF is by default 3) DN locations for each block. Example: for block1 DN2, DN1, and DN3.
    - Priority
    - Task ID
    - CPU and memory requirements

Task assignment

Task assignment is only applicable to non-uber jobs. MRAppMaster prepares a task execution plan based on the location of blocks and requests RM to allocate container for launching tasks. RM instructs YARN job scheduler to look for containers in the desired NM. Example: YARN job scheduler attempts to launch a container in NM2 to achieve data-local execution. If there is not enough resource available to launch container in NM2, the MRAppMaster requests job scheduler to allocate container in NM1 to achieve data local execution. If no data locality is possible, MRAppMaster requests job scheduler in any NM regardless of block location to go for non-local execution. Reduce tasks do not have locality constraints to satisfy but scheduled aiming to minimize network bandwidth. By default, map/reduce tasks are assigned with 1024 MB memory and 1 physical/logical core.

Task execution

9. RM instructs specific NM to start the container. For each task, NM starts a container (java process with YarnChild) with specified containerID.
10. The NM gives MRAppMaster a handle for the container.
11. YarnChild acquires and localizes job resources (configurations, jars) by creating a local working directory. Then, data blocks are brought by DN in the respective slave node.
12. YarnChild creates an instance of TaskRunner to run the task. TaskRunner launches a map/reduce task. A particular task will be attempted at least once.
13. Shuffle phase in MR is responsible for moving map outputs to the reduce node. Every NM
    runs shuffle service that retrieves map output.

Status updates

14. YarnChild provides the necessary information (task progress, counters) to its MRAppMaster via an application-specific protocol (every 3 seconds). MRAppMaster aggregates logs and counters from YarnChild to assemble the current status of the job to update clients.
15. Once the application is complete, MRAppMaster deregisters with the RM and shuts down.
16. Client (Job object) periodically polls MRAppMaster for job status update.

Job completion

Job is marked successful or failed upon its completion. RM WUI displays the details of all the running jobs and their tasks. Every job is given a URL of MRAppMaster to track job status. Because the job object does not know in which node MRAppMaster of every job is running. However, once the job is done successfully, job history information from MRAppMaster is moved to JHS. Finally, the job object displays the job counters.

5.5 PLANNING HADOOP CLUSTER

Let us discuss Hadoop cluster resource requirements, map/reduce task resource requirements, and its allocations. In this regard, basic questions that strike in our mind are:

- How many nodes do we require in a cluster for optimal data processing?
- How to plan for CPU, memory, storage, and network bandwidth for Hadoop cluster?
- When to increase the number of nodes (horizontal scalability) in the cluster?

To answer these questions, MR programers and administrators should have some minimum knowledge of Hadoop infrastructure. Let us understand in detail.

Commodity/cheap hardware

When you are operating hundreds of machines, cheap hardware turns out to be a false economy as a failure rate incurs a higher maintenance cost. On the other hand, large database-class machines are not recommended either, since they do not score well on the price/performance curve (see Figure 1.7).

Therefore, mid-range of enterprise-level servers which are commonly available in the market are highly preferred. The term commodity means that you need not depend on specific vendors for hardware such as CPU, memory, storage, and network. Hadoop works on any type of hardware from any vendor you purchase in the market.

Resource allocation strategies

There are two resource allocation strategies used in general, as shown in Figure 5.6, for task scheduling: time sharing and space sharing.

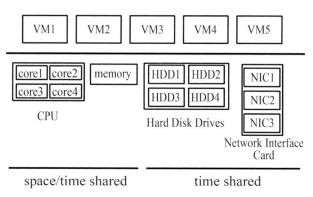

FIGURE 5.6   Resource allocation.

Space sharing (exclusive mode) – Once a task is reserved with resources, for instance, 1 GB memory and 1 core, these resources are not shared with any other tasks until its completion or termination.

Time-sharing – Allocated resources for a task is shared with other tasks in time-interleaved (round-robin) fashion.

Memory and CPU can be either space-shared or time-shared. However, IO resources are always time-shared. Map and reduce tasks are space-shared as there is no context-switch during job execution. Why? Time-sharing leads to less performance as context-switch among tasks has to move more data back and forth from virtual memory. Moreover, it is not possible to load the complete data blocks in memory, which requires substantial virtual memory to support more context-switches. Therefore, the performance of time sharing is comparatively lesser than space sharing. For data-intensive tasks, space sharing is preferred in order to minimize the average switching delay.

For compute-intensive applications, either time sharing or space sharing is good enough.

5.5.1 *STORAGE*

Hadoop deals with IO bound jobs as it has to read data from disks very fast. It is not much useful if you try Hadoop for compute-intensive jobs. The most common practice in deciding the Hadoop cluster size is based on storage capacity. Whenever you add a new node to the cluster, you get more computing resource and storage capacity. It is recommended to use HDD that supports fast read/write access, but using SSDs is a costly option. Consider a DN in the Hadoop cluster comprises two 4 TB HDD (8 TB in total). Let us quickly calculate the number of DNs required per month according to the daily input data (100 GB) of an application in the cluster.

Daily input data with default HDFS replication $p = 100$ GB x 3 = 300 GB

Input data growth per month $q = p$ x 30 = 8.78 TB

To deal MR intermediate data $r = 15\%$ per disk

Non-HDFS storage requirement $s = 15\%$ per disk

HDFS data space per disk $= 4$ TB x (1-r-s) = 4 TB x 0.7

$= 2.8$ TB per disk

Per node capacity $= 2$ x 2.8 TB = 5.6 TB

Number of nodes required per month $= q / 5.6 = 1.56$ nodes

How about using SSDs? SSDs are highly preferred for compute-intensive applications. SSDs contain no rotational components that are fragile and noisy. But it is too costly. In the future, we will be using SSD in Hadoop cluster.

RAID vs replication

RAID ensures that there is no data loss despite disk failures. RAID1 is just mirroring (also called replication). If there are two 1 TB HDD in RAID array, whatever is written in the first drive is replicated to the second drive. This is called mirroring and mainly useful in NN and RM as there is no replication concept used for maintaining their meta-data. In contrast, the HDFS itself is designed to make the data replication instead of hardware-level replication like RAID.

Why should not we use RAID for slaves?

The HDFS already handles data loss with the help of replication. RAID might be useful for NN and RM to handle SPOF. However, using RAIDs for salves is again similar to centralized storage that requires movement of blocks across the local network during job execution. The primary requirement of MR jobs is to achieve data-local execution to minimize local network band-width consumption, which is not possible in RAID-based storage schemes in the Hadoop cluster. So, replication provided by HDFS is the better option to distribute data blocks across the cluster for smooth parallel processing. There are some drawbacks of RAID for Hadoop cluster:

- Slowest disk read/write performance in a RAID array decides the entire RAID read/write performance. If RAID contains three disks whose performance are: disk1 50 MB/s, disk2 10 MB/s, and disk3 50 MB/s, then the performance of this RAID rack is 10 MB/s.
- If a disk in RAID failed, the entire RAID array is considered to be failed.

Therefore, Just in Bunch Of Disks (JBOD) is preferred over RAID. JBOD is a set of HDD connected to a node. The total performance of JBOD is an average of all HDD's performance. Example: if there are three HDDs attached to a node whose performance are: 50 MB/s, 10 MB/s, and 50 MB/s, then the performance of JBOD is 50+10+50=110/3=37MB/s. If one HDD is failed in a JBOD, it is easy and affordable to replace it with a new HDD. Performance of one 6 TB HDD in a node is lesser than using three 2 TB HDDs. Therefore, it is recommended to add several smaller size HDDs than using less number of high capacity HDDs. Because many map/reduce tasks can read/write data from different disks simultaneously instead of being served by single disk linearly.

Buffer size

For all IO operations, the reserved buffer size is 4 KB. You can observe performance increment if you increase the buffer size to 128 KB. This IO buffer size for MR execution process can be set by using io.file.buffer.size property in core-site.xml.

Reserved storage space

Storage in every DN is divided into HDFS and non-HDFS parts. The HDFS extends its storage capability by taking space from the non-HDFS portion.

However, you can reserve space for non-HDFS in disks by using dfs. datanode.du.reserved property.

Trash

Deleted files in HDFS are temporarily kept in the trash for a specified amount of time instead of removing completely. Time (in minutes) can be customized using fs.trash.interval property in core-site.xml. Trash is disabled by default (which means 0). Files deleted via external programming are deleted without keeping them in the trash. When the trash is enabled, each Hadoop user has its trash directories.

5.5.2 CPU

The desktop/laptop is not designed to run 24/7 without interruption. Therefore, Hadoop nodes should have server CPU such as Xeon/AMD processor to form Hadoop physical cluster. Before proceeding further, let us understand what physical core, logical core, and virtual core are. In general, a physical core denotes an ALU in the processor that handles one thread. If there are four ALU in a processor, it is called a quad-core processor. If Hyper-Thread (HT) is enabled, then a physical core can execute two threads at the same time. So, a physical core represents two logical cores based on the number of threads it can execute at the same time. HT shows a physical core to the OS as if there are two real physical cores available. So, OS provides an illusion to Hadoop services that two cores are available instead of one core. Now, the YARN job scheduler assigns two tasks to the same core. However, in reality, these two tasks are concurrently executed, but in super-fast. Therefore, an HT core is utilized thoroughly (100%). We can set the HT utilization factor between 95-175%.

If HT supports four threads, then a physical core is equivalent to four logical cores. If a processor has 4 physical cores with HT (two-threads), then there are 8 logical cores available which can serve physical 8 threads (map/ reduce tasks) simultaneously. In virtualization, a physical core assigned to a VM is called a virtual core. Hadoop daemons and map/reduce tasks are assigned with 1 physical core (or logical core if HT enabled) for better performance as they are space shared. Let us assume that a slave node has 8 logical cores. To find the number of map/reduce tasks possible in that slave node,

= (# logical cores – # reserved logical cores) x (a value between 0.95 to 1.75)

Reserved logical cores = DN + NM + OS = 1 + 1 + 1= 3 logical cores

Assume each core performance is up to 150%. Therefore, the maximum number of map/reduce tasks that can be launched is = (8 – 3) x 1.5 = 7.5. So, up to 8 map/reduce tasks per node can be executed. Allowing 1-2 containers per disk and per core gives a better balance for cluster utilization. So, a node with 12 disks and 12 cores can comfortably deal with 24 map/reduce tasks.

5.5.3 MEMORY

It is easy to calculate the memory needed for both NN and SNN. The memory needed for NN is determined based on the number of blocks stored in HDFS. Typically, the memory needed by SNN should be identical to NN. Example:

Memory needed for NN = meta-data + NN daemon + OS = meta-data size + 1 GB + 2 GB

Amount of memory required for the slave node is dependent on how many logical cores possible in a node. Consider 2 HT dual-core processors (8 logical cores in total) in a slave node. DN, NM, and OS take 1 logical core each. Therefore, only 5 map/reduce tasks can be executed in parallel. According to this, a slave node memory requirement is,

= DN daemon + NM daemon + OS + (map/reduce task memory in average x # cores)

= 1 GB + 1 GB + 2 GB + (1 GB x 5) = 9 GB RAM per slave node

If the number of logical cores is 4, despite having 32 GB memory, it is possible to launch only 2-4 map/reduce tasks. Therefore, there is no use of having a huge memory without equivalent CPU capacity. Deploying Hadoop cluster in the virtual environment requires 1 GB memory and 1 logical core for the hypervisor. If 5 VMs are hosted in a server, the server should have minimum

- 45 GB memory (each slave nodes take 9 GB as calculated above)
- 4 quad core HT processors
- 8 x 2 TB HDD
- three 1 Gbps NIC

5.5.4   NETWORK BANDWIDTH

Hadoop cluster traffic is high during data loading, replication, interme-diate data transfer, and other such tasks. Every node has more than one NIC (1 Gbps), so data can be transferred simultaneously. If one NIC fails, the other one can be used. Hadoop cluster should have dedicated switches. HTTPServlet protocol is used in the shuffling process to transfer data over the local network. Every node in Hadoop cluster runs (i) Remote Procedure Call (RPC) server and (ii) HTTP server to serve web pages and transfer data over the network. DNs additionally run TCP/IP server to transfer blocks.

5.5.5   CONTAINER VS JVM

Is container the same as JVM? It is an inevitable confusion while talking about resource reservation. The JVM is a program running in memory to provide a platform-independent execution environment for Java programs by converting Java bytecode into machine codes that are suitable to target platforms. The container is a logical pack of memory and logical cores. Therefore, JVM and container are not the same. The JVM runs in a container, taking up memory and logical core available in the container. More precisely, JVM is a subset of the container. However, JVM has the rights to occupy the entire memory and cores available in the container.

As shown in Figure 5.7, assume a node with 4 GB memory and 4 cores. A container occupies 1 GB memory and 1 core. Now, a JVM running in the can occupy a maximum 1 GB memory and 1 core. A JVM consists of the following segments.

- Heap Memory: data blocks, intermediate data, runtime MR job objects, etc. are loaded.
- Non-Heap Memory: supporting jars, execution environment jars to convert byte code to machine code is loaded into this area.

Heap size is configured with the following options:

- Xms<size> sets the initial heap size.
- Xmx<size> sets the maximum heap size in JVM.

Example: To run a java program with heap size (256 MB to 2048 MB) in JVM

$ java -Xmx2048m -Xms256m classname

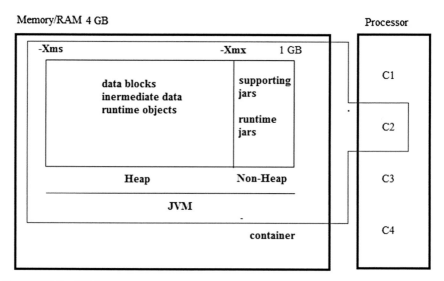

FIGURE 5.7   JVM.

Memory flag (m, k, g) indicates different sizes (kilobytes, megabytes, gigabytes, etc.). For example, -Xmx1024k or -Xmx512m or -Xmx8g. Xmx flag has no default value, but Xms typically holds default size 256 MB. If you encounter a java.lang.OutOfMemoryError, it means that Heap memory is not sufficient for the task. The default value of Xms is dependent on the platform and the amount of memory available in the system. It is your responsibility to configure it correctly, so it uses the available memory. In general, TT/NM kills any task that demands more than given heap size. However, this can be configured in NM to add more memory if a task requires.

There are many factors which can affect the memory required for a container. Such factors include file type (plain text file, parquet, ORC), data compression algorithm, type of operations (sort, group-by, aggregation, join). You should be familiar with the nature of MR jobs and figure out the minimum requirement of the heap for tasks. Example: Parquet is a column-oriented binary file format. If an MR job sorts parquet files, the entire column has to be loaded in memory. Its column size may be varying.

So, we need to adjust the JVM size. To set 1024 MB heap memory for tasks, use -Dmapred.child.java.opts='-Xmx1024m' while launching MR job via command line or include the following piece of code in the MR driver program.

```
conf.set("mapred.child.java.opts", "-Xmx1024m")
```

We will discuss these options in the subsequent sections.

5.6   MEMORY AND CPU CONFIGURATION

Hadoop daemons take 1 GB memory and 1 logical core to run in the background. This can be customized in hadoop-env.sh file. By default, the YARN scheduler allocates containers with 1 GB memory, and 1 logical core for map/reduce tasks. We can also customize how much amount of memory should be allocated for map/reduce tasks using MRv2 job configuration properties to improve performance. There are two primary configuration files in YARN: yarn-site.xml and mapred-site.xml. Cluster, RM, NM, and container-related configurations are mentioned in yarn-site. xml. Map/reduce tasks, and JVM-related configurations are set up in the mapred-site.xml file.

Before configuring memory and CPU, you must know the amount of memory, and the number of logical cores available in an NM. As shown in Figure 5.8, YARN properties start with the keyword "yarn." Example: yarn.nodemanager.resource.memory.mb (default 8192 MB). It conveys the amount of memory allocated for NM in that slave. The MR properties start with the keyword "mapreduce." Example: mapreduce.map.memory. mb. The property yarn.scheduler. maximum-allocation-mb (default 8192 MB) and yarn.scheduler.minimum-allocation-mb (default 1024 MB) restrict the maximum and minimum allocation of memory for a container by the scheduler. MRAppMaster can request RM additional resources for map/ reduce tasks with an increment of yarn.scheduler.minimum-allocation-mb and do not exceed yarn. scheduler.maximum-allocation-mb. mapreduce. map.memory.mb and mapreduce.reduce.memory.mb decide how much amount of memory from container can be allocated for map and reduce tasks. However, it should not exceed yarn.scheduler.maximum-allocation-mb, and minimum should be yarn.scheduler. minimum-allocation-mb. It should be capable enough to load intermediate job data to avoid too many spills involving more IO operations. mapreduce.map.java.opts and mapreduce. reduce.java.opts decide how much memory should be given to JVM

Heap portion. However, it should not exceed mapreduce.map.memory.mb and mapreduce. reduce.memory.mb. In general, heap size should be 80% of memory allocated to map/reduce tasks. If a task runs out of heap memory, an exception (java.lang.RuntimeException:java.lang. OutOf MemoryError) is thrown.

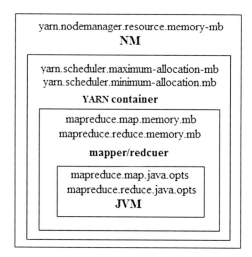

FIGURE 5.8   Resource allocation.

Containers are launched in NMs only when yarn.nodemanager.disk-health-checker.min-healthy-disks (default value 0.25) is true. The increasing memory size of map/reduce tasks comes at the expense of a reduction in parallelism as it minimizes the number of containers possible in a node. So, understand your data at every step in the MR execution sequence and configure CPU, memory for containers. For instance, if a node contains 16 GB memory, 1 TB HDD and 8 logical cores, memory is allocated for NM, map/reduce tasks, etc. as shown in Figure 5.9. OS takes a minimum of 2 GB, and NM captures 14 GB from 16 GB. Hadoop services such as NM and DN require 1024 MB each for its JVM. So, NM has 12 GB for launching containers.

YARN scheduler settings

yarn.scheduler.minimum-allocation-mb = 2048 MB
yarn.scheduler.maximum-allocation-mb = 4096 MB

MR settings

mapreduce.map/reduce.memory.mb = 2048 MB (must be greater than or equal to yarn. scheduler.minimum-allocation-mb and must be less than yarn.scheduler.maximum-allocation -mb). If MRAppMaster requests additional resources for a container, the scheduler will allocate in increments of yarn.scheduler.minimum-allocation-mb. It can reach up to yarn. scheduler. maximum-allocation-mb. JVM heap size (mapreduce.map/reduce.java.opts) is set to 80% of mapreduce.map/reduce.memory. mb. If map/reduce task's heap usage exceeds 1.6 GB, NM will kill the task if there is no request to additional memory. The MRAppMaster will reschedule the task. Virtual memory allocation is limited by NM. A task is killed if it exceeds its allowed virtual memory. It is configured in multiples of physical memory.

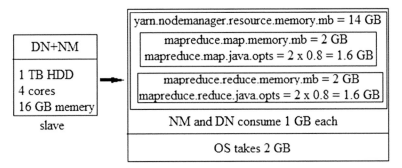

FIGURE 5.9 YARN memory configuration.

yarn.nodemanager.vmem-pmem-ratio is set to 2.1 by default. This means that a map/reduce container can allocate virtual memory up to 2.1 times of mapreduce.map/reduce.memory.mb before NM kills the task. If mapreduce.map/reduce.memory.mb is set to 2048, the total allowed virtual memory is 2.1 *2048 = 4300.8 MB. You will see several counters dumped at the end of job execution in the console. Memory counters show statistics of memory allocation: physical memory (bytes), virtual memory (bytes), total committed heap usage (bytes). Table 5.3 gives you the default values for important configuration properties in mapred-site.xml and yarn-site.xml. You can override these values at run time.

TABLE 5.3 Default Values for Important YARN and MR Properties

yarn-site.xml	Default value
yarn.scheduler.minimum-allocation-mb	1024
yarn.scheduler.maximum-allocation-mb	8192
yarn.nodemanager.vmem-pmem-ratio	2.1
yarn.scheduler.minimum-allocation-vcores	1
yarn.scheduler.maximum-allocation-vcores	32
mapred-site.xml	**Default value**
mapreduce.map.memory.mb	1024
mapreduce.map.java.opts	-Xmx900m
mapreduce.map.cpu.vcores	1
mapreduce.map.disk	0.5
mapreduce.reduce.memory.mb	1024
mapreduce.reduce.java.opts	-Xmx900m
mapreduce.reduce.cpu.vcores	1
mapreduce.reduce.disk	1.33
yarn.app.mapreduce.am.resource.mb	1024
yarn.app.mapreduce.am.command-opts	-Xmx900m
yarn.app.mapreduce.am.resource.cpu-vcores	1

Key to remember:

- Most of the job failures in cluster mode are due to improper YARN, and MR configuration.
- The container is a package of CPU and memory. JVM can be said as a subset of the container.
- Heap portion in JVM loads data blocks, intermediate data, etc.
- yarn.nodemanager.resource.memory.mb sets the amount of memory for NMs.
- Values of mapreduce.map/reduce.memory.mb should be at least yarn.scheduler. minimum-allocation-mb and maximum up to yarn. scheduler.maximum-allocation-mb.
- Values of mapreduce.map/reduce.java.opts should be around 80% of mapreduce. map/reduce.memory.mb.
- Value of yarn.app.mapreduce.am.command-opts should be 80% of yarn.app.map reduce.am.resource.mb.
- If the block size is 128 MB, Heap memory should be at least 512 MB.

- We cannot determine the number of containers possible in an NM as container size of each job can be configured differently.
- Virtual memory for map/reduce should be 2.1 times of map/reduce memory.

5.6.1   RESOURCE REQUEST MODEL

Every submitted job gets an MRAppMaster, which requests RM to allocate containers for its map/reduce tasks or additional resources for the running tasks. This request consists of a sequence of steps and carries some information, as shown in Figure 5.10. The steps are,

1. MRAppMaster requests RM for containers to launch map/reduce task or additional resources for containers. Resource request by MRAppMaster comprises:
   <resource-name, priority, resource-requirement, number-of-containers>
2. RM gives control to the scheduler, which launches containers in the requested NM.
3. NM launches container and reports back to RM.
4. RM gives control to MRAppMaster with node address and container ID.
5. MRAppMaster configures and manages the job life cycle.

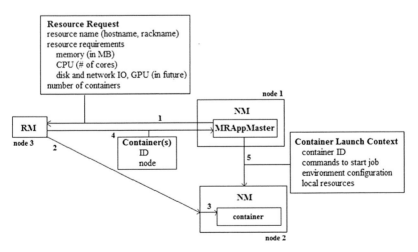

FIGURE 5.10   Container allocation.

5.7   MRV2 FAILURE HANDLING

There is a chance of failure in map/reduce tasks, MRAppMaster, NM, RM, as shown in Figure 5.11. Hadoop handles all these failures at the software-level and leads jobs to complete successfully without any extra coding effort from the programer. Unlike MRv1, failure handling responsibility is split and given to different services.

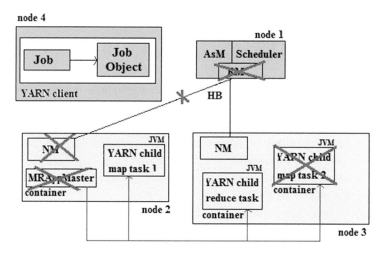

FIGURE 5.11   Failure cases in YARN.

Task failure

A map/reduce task may fail due to various reasons such as exception (due to buggy code), JVM crash, hardware or OS crash, memory over-consumption. So, the NM kills such tasks and removes its container. MRAppMaster comes to know when YarnChild stops sending task progress. It marks task attempt as failed and will re-attempt the failed tasks by contacting RM. This time, MRAppMaster schedules the failed task in some other NM. Jobs with too many failed tasks are considered to be failed. Hanging (infinite loops) tasks are killed, and task attempt is marked as failed. This can be customized via mapreduce.task.timeout (default 10 mins) property. Setting zero for this property will disable time out and never marks the tasks as failed. A job is considered to be failed if any one of the tasks fails for four times. You can modify this with mapreduce.map.maxattempts, and mapreduce.reduce.maxattempts. Sometimes, even if a few tasks failed, the final result of a job may be useful. So, we can set the maximum percentage of tasks that are allowed to fail without triggering job failure using mapreduce.map/reduce. failures.maxpercent.

MRAppMaster failure

When MRAppMaster fails, it stops sending HB to AsM. This results in the whole job are considered to be failed. Because all tasks running under the job are killed and their containers are removed. The MRAppMaster is re-executed by AsM using resourcemanager.am.max.retries (default 2) property. Therefore, the whole job is re-executed from the beginning. If the job is too big, re-executing all the tasks (including task completed in the last attempt) is time-consuming. Therefore, MRAppMaster has yarn.app.mapreduce.am.job. recovery.enable (true by default) property for job recovery. If it is false, all tasks will be re-run. If it is true, only the killed or incomplete tasks are re-executed.

NM failure

If there is no HB from NM to RM for 10 mins (default), RM will remove NM from the cluster. The NM may fail due to OS/hardware crash or running very slowly.

- Tasks running in the failed NM will be treated as failed by MRApp-Master and rescheduled to some other NM.
- Successful map tasks are rescheduled to different NM as map output cannot be recovered from the crashed NM. Successful reduce tasks need not be re-executed as its output is replicated to multiple DNs.
- The MRAppMaster running in the failed NM also is considered to be failed and re-executed by AsM.

You can configure NM expiration with yarn.resourcemanager.nm. liveness-monitor.expiry interval-ms. If 4 or more tasks from the same job failed in a particular NM, RM considers such nodes as faulty NM. When the minimum threshold of faults exceeded, NM is blacklisted by RM even if NM is not crashed. However, faults expire over time (one per day), and NM gets a chance to run jobs again. The MRAppMaster also adds NM in blacklist per job basis. Therefore, it does not affect other jobs.

RM failure

RM failure is catastrophic as no jobs can be launched. All running applications are terminated if the RM fails. RM also suffers from SPOF. Although the probability of RM failure is relatively low, no jobs can be submitted until

RM is brought back to a running condition. HA can be configured by running multiple RM in the cluster. There is no HA for JT in MRv1. However, you can save state by configuring yarn.resourcemanager.store.class property. After RM restart, all jobs must be re-submitted.

Speculative execution

A job is not completed until all of its map/reduce tasks are finished. So, if there is a long-running task (map/reduce) without much progress, overall job latency is dragged to be high. Such a task is called a straggler. In Figure 5.12, map task5 in NM2 takes so much time causing overall job latency high. Generally, a Hadoop cluster may be formed with heterogeneous nodes.

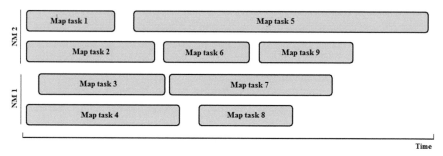

FIGURE 5.12   Job latency.

As a result, there can be very slow nodes as well as fast nodes. Potentially, a few slow nodes can slow down task execution significantly. Software misconfiguration or hardware degradation can cause a task to progress slower than peer tasks. Therefore, the overall job progress is delayed. So, the MRAppMaster launches a redundant task for stragglers in some other NM to finish it faster. If the original task finishes first, the redundant task will be killed and vice versa. This is called speculative execution. The major drawback of speculative execution is that the resource consumption for a task will be high as a copy of the same task is executed simultaneously along with the original task.

These impacts other jobs in the cluster. It is mainly for hardware issues and not meant for buggy code. Speculation is enabled for both map and reduce tasks. You can disable it if the cluster is heavily loaded, by setting mapreduce.map/reduce.speculative to false. Enabling speculation for reduce tasks consumes more bandwidth as a particular reducer needs to bring its input from all completed map tasks. Therefore, the output of map tasks is not deleted until all reduce tasks to receive their inputs or for a time limit. Speculative tasks are launched only after all the tasks of a job have been launched. It is an optimization feature, not a reliability feature. It means that it does not guarantee to fix the coding bug.

Reduce slow start

Reduce function is executed only after the inputs from all map tasks are received. However, the map output collection process (reduce phase) is started only after a certain number of map tasks are completed. Schedulers wait until 5% of map tasks get over before reduce phase starts. For large jobs, this can cause problems with cluster utilization because they hold reduce containers while waiting for the map tasks to complete. Setting mapreduce. job.reduce. slowstart.completedmaps to a higher value, 0.8 (80%), can help improve throughput.

5.8 JVM REUSE

A container generally runs a map/reduce task. Forking and scheduling containers for jobs having short-lived tasks add more overhead than its latency. Therefore, it is possible to run more than one map/reduce tasks in a JVM. Example: If we assign four map tasks per JVM, the first task gets a container, which is reused by serially executing the rest of the three tasks. This minimizes the latency of a job. By default, JVM is set to run 1 task (mapreduce.job.jvm. numtasks=1). When to enable JVM reuse? When you infer a job with a large number of small and short-lived map/reduce tasks, you can set it to greater than 1. It is your responsibility to ascertain the duration of tasks and go for JVM reuse.

Uber job

When map/reduce tasks of a small job is scheduled into MRAppMaster JVM itself, it is called uber job, and the tasks are called uber tasks. Uber mode is not enabled by default. To enable uber mode, set mapreduce.job.uber-task.enable to true. The choice is done by judging the overhead of JVM creations. If the job is bigger, tasks are executed in their containers. The MRAppMaster will uberize the job if all the following conditions are met:

- A number of map tasks must be <= 9 (mapreduce.job.ubertask. maxmaps).

- A number of reduce task should be 1 (mapreduce.job.ubertask. maxreduces).
- Input size should be less than 1 HDFS block (mapreduce.job.ubertask.maxbytes).

You can override the above properties, but only downward. While running uber jobs, MR framework disables speculative execution and also sets the maximum attempts for the task to 1.

KEYWORDS

- **container in MapReduce v2**
- **failure handling**
- **Hadoop 2.x components**
- **hardware and container configuration in MapReduce v2**
- **YARN**

CHAPTER 6

Hadoop 2.7.0 Installation

All of us do not have equal talent but,
all of us have an equal opportunity to develop our talents.
—Abdul Kalam

INTRODUCTION

Please refer Section 3.1 for system requirements to install Hadoop. Chapter 3 discusses in detail about single node, multi-node installation and other basic requirements for installation. In this chapter, we have given only commands for the installation, not explanation. So, we recommend you to refer Hadoop 1.x installation side-by-side to understand every step in detail.

6.1 SINGLE NODE SETUP

6.1.1 STANDALONE OR LOCAL MODE (REFER SECTION 3.2.1)

Step 1: Download Hadoop 2.7.0 and unpack the tar file

https://archive.apache.org/dist/hadoop/core/

$ wget https://archive.apache.org/dist/hadoop/core/hadoop-2.7.0/hadoop-2.7.0.tar.gz

$ tar -zxvf hadoop-2.7.0.tar.gz

Step 2: Move Hadoop file to /usr/local/hadoop

$ sudo cp -r hadoop-2.7.0 /usr/local/hadoop

Step 3: Set path to Hadoop

$ vi .bashrc

```
export HADOOP_HOME=/usr/local/hadoop
export PATH=$PATH:$HADOOP_HOME/bin
export PATH=$PATH:$HADOOP_HOME/sbin
export HADOOP_HDFS_HOME=$HADOOP_HOME
export HADOOP_MAPRED_HOME=$HADOOP_HOME
export HADOOP_COMMON_HOME=$HADOOP_HOME
export YARN_HOME=$HADOOP_HOME
export HADOOP_CONF_DIR=$HADOOP_HOME/etc/hadoop
export HADOOP_COMMON_LIB_NATIVE_DIR=$HADOOP_HOME/
lib/native
export HADOOP_OPTS="-Djava.library.
path=$HADOOP_HOME/lib"
```

$ source .bashrc

$ $PATH

Step 4: Configure Hadoop environment variable

$ sudo vi /usr/local/hadoop/etc/hadoop/hadoop-env.sh

```
export JAVA_HOME=/usr/lib/jvm/java-8-openjdk-amd64
export HADOOP_OPTS=-Djava.net.preferIPv4Stack=true
```

$ hadoop version or yarn version

```
Hadoop 2.7.0
...
Compiled by jenkins on 2015-04-10T18:40Z
Compiled with protoc 2.5.0
From source with checksum
a9e90912c37a35c3195d23951fd18f
This command was run using /usr/local/hadoop/share/
hadoop/common/hadoop-common-2.7.0.jar
```

Step 5: Grant privileges

Grant all access rights to the Hadoop daemons to read and write from a local file system location /usr/local/hadoop. Username is **itadmin.**

$ sudo chown -R itadmin /usr/local/hadoop

$ sudo chmod -R 777 /usr/local/hadoop

Step 6: Run a simple wordcount job

No MR and HDFS daemons are running in this mode. Create a file "input. txt" in the local file system and launch MR wordcount job with output directory "result," which is created in the local file system, not in HDFS. Output directory should not already exist. Once the job is submitted, LocalJob-Runner will take care of the MR job execution sequence as there is no MR daemon running.

```
$ vi input.txt
    hi how are you
$ yarn jar /usr/local/hadoop/share/hadoop/mapreduce/hadoop-
mapreduce-examples-2.7.0.jar wordcount input.txt result
                            // result is the output directory name
$ ls result
    part-r-00000
    _SUCCESS
$ cat result/part-r-00000
    are      1
    hi       1
    how      1
    you      1
```

6.1.2   PSEUDO-DISTRIBUTED MODE

Assume IP of the working node (PM/VM) is **10.100.55.92** and username is **itadmin.** Installation in short,

1. Generate a public-private key pair for passwordless communication.
2. Download Hadoop 2.7.0 and untar.
3. Move Hadoop file to /usr/local/hadoop location.
4. Set path to Hadoop in .bashrc file.
5. Configure Java_Home, disable IPv6 in hadoop-env.sh
6. Edit configuration files.
7. Change ownership, access mode, and create Namespace.
8. Start HDFS/YARN.
9. Run a simple wordcount program.

Step 1: Generate public-private key pair for passwordless communication

> $ ssh-keygen or ssh-keygen -t rsa   // leave empty for name, password
> $ cat ~/.ssh/id_rsa.pub >> ~/.ssh/authorized_keys
> $ chmod 0700 ~/.ssh/authorized_keys

Step 2: Download Hadoop2.7.0 and unpack the tar file

> https://archive.apache.org/dist/hadoop/core/
> $ wget https://archive.apache.org/dist/hadoop/core/hadoop-2.7.0/hadoop-2.7.0.tar.gz
> $ tar -zxvf hadoop-2.7.0.tar.gz

Step 3: Move Hadoop file to /usr/local/hadoop

> $ sudo cp -r hadoop-2.7.0 /usr/local/hadoop

Step 4: Set path to Hadoop in .bashrc file

> $ vi .bashrc

```
export HADOOP_HOME=/usr/local/hadoop
export PATH=$PATH:$HADOOP_HOME/bin
export PATH=$PATH:$HADOOP_HOME/sbin
export HADOOP_HDFS_HOME=$HADOOP_HOME
export HADOOP_MAPRED_HOME=$HADOOP_HOME
export HADOOP_COMMON_HOME=$HADOOP_HOME
export YARN_HOME=$HADOOP_HOME
export HADOOP_CONF_DIR=$HADOOP_HOME/etc/hadoop
export HADOOP_COMMON_LIB_NATIVE_DIR=$HADOOP_HOME/
lib/native
export HADOOP_OPTS="-Djava.library.
path=$HADOOP_HOME/lib"
```

> $ source .bashrc
> $ $PATH

Step 5: Configure Hadoop environment variable

> $ sudo vi /usr/local/hadoop/etc/hadoop/hadoop-env.sh

```
export JAVA_HOME=/usr/lib/jvm/java-8-openjdk-amd64
export HADOOP_OPTS=-Djava.net.preferIPv4Stack=true
```

> $ yarn version

Step 6: Edit configuration files

To set up NN

$ sudo vi /usr/local/hadoop/etc/hadoop/core-site.xml

```
<configuration>
    <property>
        <name>fs.defaultFS</name>
        <value>hdfs://10.100.55.92:10001</value>
    </property>
    <property>
        <name>hadoop.tmp.dir</name>
        <value>/usr/local/hadoop/tmp</value>
    </property>
</configuration>
```

fs.defaultFS – specifies IP address of NN
hadoop.tmp.dir – specifies the local file system location for storing FSImage, edit logs in NN, SNN, meta-data of JT, spilling map output, to store blocks in DNs.

To set up YARN

$ sudo vi /usr/local/hadoop/etc/hadoop/yarn-site.xml

```
<configuration>
    <property>
        <name>yarn.nodemanager.aux-services</name>
        <value>mapreduce_shuffle</value>
    </property>
    <property>
        <name>yarn.nodemanager.aux-
        services.mapreduce.shuffle.class</name>
        <value>org.apache.hadoop.mapred.
        ShuffleHandler</value>
    </property>
</configuration>
```

In MR1, TT sends intermediate data to the reducer. However, in MR2, NM does not do that (because it is a YARN component). So, we have different auxiliary services that must be set in yarn-site.xml.

yarn.nodemanager.aux-services – Containers do not know what is being executed inside since YARN may run any data processing tool's job. So, map tasks running inside the container do not know how to perform shuffle. Therefore, this property explicitly specifies that shuffle should be taken care of by the MR framework itself.

yarn.nodemanager.aux-services.mapreduce.shuffle.class – denotes the library class that handles shuffle and sort.

To set up MR-related properties

$ sudo cp /usr/local/hadoop/etc/hadoop/mapred-site.xml.template / usr/local/hadoop/etc/hadoop/mapred-site.xml
$ sudo vi /usr/local/hadoop/etc/hadoop/mapred-site.xml

```
<configuration>
    <property>
        <name>mapreduce.framework.name</name>
        <value>yarn</value>
    </property>
</configuration>
```

mapreduce.framework.name – specifies the resource management component as YARN.

To set up NN, SNN, DN configuration (optional)

$ sudo vi /usr/local/hadoop/etc/hadoop/hdfs-site.xml

```
<configuration>
    <property>
        <name>dfs.replication</name>
        <value>3</value>
    </property>
</configuration>
```

To set up SNN: SNN is automatically runs with NN, however, you can set up.

$ sudo vi /usr/local/hadoop/etc/hadoop/masters

```
10.100.55.92
```

To set up slave (DN and NN)

$ sudo vi /usr/local/hadoop/etc/hadoop/slaves

```
10.100.55.92
```

Step 7: Change ownership, access mode, and create Namespace

$ sudo chown -R itadmin /usr/local/hadoop

$ sudo chmod -R 777 /usr/local/hadoop

$ hdfs namenode -format   // HDFS namespace is created

$ ls /usr/local/hadoop/tmp/dfs/name/current // you see something as follows

```
fsimage_0000000000000000000
fsimage_0000000000000000000.md5 seen_txid VERSION
```

Step 8: Start HDFS/YARN (3 ways)

Way 1: To start/stop dfs and yarn

$ start-dfs.sh   // location for DN and SNN to store are created

$ ls /usr/local/hadoop/tmp/dfs/

```
    name data secondarynamenode
```

$ jps

```
    29557 datanode
    29753 secondarynamenode
    29403 namenode
```

$ start-yarn.sh

$ ls /usr/local/hadoop/tmp/   // location for RM and NM to store are
                                created

```
    dfs nm-local-dir
```

$ jps

```
    30259 NodeManager
    29557 DataNode
    29753 SecondaryNameNode
    29403 NameNode
    29949 ResourceManager
```

$ stop-yarn.sh

Way 2: To start/stop services altogether

$ start-all.sh

$ jps

$ stop-all.sh

Way 3: To start/stop services individually

To start/stop NN: $ hadoop-daemon.sh start/stop namenode
To start/stop DN: $ hadoop-daemon.sh start/stop datanode
To start/stop SNN: $ hadoop-daemon.sh start/stop secondarynamenode
To start/stop YARN:$ yarn-daemon.sh start/stop resourcemanager
To start/stop NM: $ yarn-daemon.sh start/stop nodemanager
                                $ jps

To make sure all services are running
    $ hdfs dfsadmin -report   // displays HDFS information
    $ hdfs fsck / -blocks -files -locations // displays block-level information
    $ yarn node -list   // displays NM information

Step 9: Run a wordcount program

    $ hdfs dfs -mkdir /data
    $ vi input.txt   // enter some text
    $ hdfs dfs -copyFromLocal input.txt /data
    $ hdfs dfs -ls /input
    $ yarn jar /usr/local/hadoop/share/hadoop/mapreduce/hadoop-
    mapreduce-examples-2.7.0.jar wordcount /data/input.txt /output
    $ hdfs dfs -ls /output
    $ hdfs dfs -cat /output/part-r-00000
    $ hdfs dfs -ls file:///   // lists out local file system root
    $ hdfs dfs -ls hdfs://10.100.55.92:10001/   // lists out HDFS meta-data

Validating output (refer Section 3.2.2.4)

- WUI
- HDFS commands
- MR and HDFS API in java programs

1. In Hadoop 2.7.0, HDFS web portal looks different than the older version. For example, browse using NN IP and port number, http://10.100.55.92:50070/ (Figure 6.1)

    Go to utilities → browse the file system to see the input and output files.

2. Web UI for application/job details: http://10.100.55.92:8088/ (Figure 6.2)

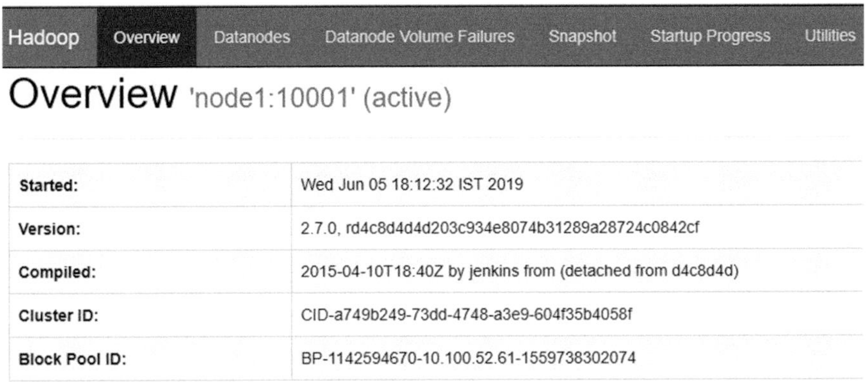

FIGURE 6.1   NN WUI.

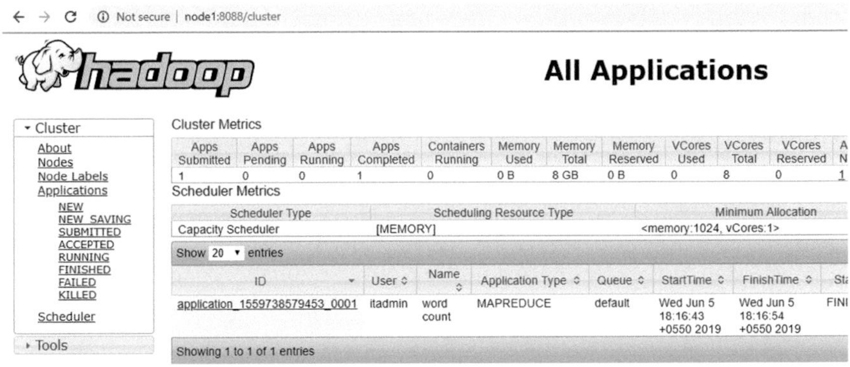

FIGURE 6.2   WUI for applications.

6.2   MULTI-NODE SETUP (CLUSTER OR FULLY DISTRIBUTED MODE DEPLOYMENT)

Follow the requirements given in Section 3.1 for all the nodes in the cluster before beginning installation. Refer Figure 3.5 and Figure 5.3 to visualize how Hadoop 2.x multi-node is set up on a cluster of PMs and a cluster of VMs. Refer Tables 3.2 and 3.3 to know the possible cluster configuration for Hadoop. Consider five PMs/VMs as given in Table 6.1. Most of us would not have multiple PMs to experiment fully distributed mode. So, we can create 5 VMs and assign IP as given in Table 6.1 for multi-node installation. You must have at least a quad-core processor, 32 GB memory, and over 500 GB HDD to launch 5 VM instances in a

PM. IP, hostname, domain name in Table 6.1 are based on my cluster. Here, the term "node" means either PM or VM. You have to fill this table according to your cluster configurations. If it is possible to launch only two VMs, then you can configure NN+RM+JHS+SNN in the first VM and DN+NM can be configured in second VM. You can launch multiple VMs in different PMs also.

TABLE 6.1   Multi-Node Implementation

VM/PM	Daemon	IP	Hostname	Domain name
node1	NN	10.100.55.92	ubuntu1	node1
node2	RM+JHS	10.100.55.93	ubuntu2	node2
node3	SNN	10.100.55.94	ubuntu3	node3
node4	Slave1 (DN+RM)	10.100.55.95	ubuntu4	node4
node5	Slave2 (DN+RM)	10.100.55.96	ubuntu5	node5

6.2.1   OS TYPE AND JAVA VERSION IN ALL PHYSICAL/VIRTUAL NODES

1. OS type must be same in all nodes (either 32-bit OS or 64-bit OS, but not mixed up). Use the following command to check OS type. If it is not the same, it is better to install the right version.

    $ uname -a

    Result for 32-bit Ubuntu will be as below:

    ```
    Linux discworld 2.6.38-8-generic #42-Ubuntu SMP Mon
    Apr 11 03:31:50 UTC 2011 i686 i686 i386 GNU/Linux
    ```

    Result for 64-bit Ubuntu will be as below:

    ```
    Linux ubuntu 4.4.0-104-generic #127-Ubuntu SMP
    Mon Dec 11 12:16:42 UTC 2017 x86_64 x86_64 x86_64
    GNU/Linux
    ```

    Alternatively, you can use the following command

    $ uname -i

    ```
    x86_64
    ```

2. Install Java in all the nodes with the same version (OpenJDK/Oracle JDK) and architecture (32-bit/64-bit). Refer Section 3.1 to install Java. To check java version and architecture,

```
$ java -version
    openjdk version "1.8.0_131"
    OpenJDK Runtime Environment (build
    1.8.0_131-8u131-b11-2ubuntu1.16.04.3-b11)
    OpenJDK 64-Bit Server VM (build 25.131-b11,
    mixed mode)
$ javac -version
    javac 1.8.0_131
```

or go to /usr/lib/jvm

```
$ ls /usr/lib/jvm
```

Result for 32-bit java will be as below:

```
    java-1.8.0-openjdk-i386 java-8-openjdk-i386
```

Result for 64-bit Ubuntu:

```
    java-1.8.0-openjdk-amd64 java-8-openjdk-amd64
```

Caution:

1. Username must be the same in all the nodes. I have set "**itadmin**" as the username for all the nodes in my cluster. So, create a common username in all nodes in the cluster using the following command.

    `$ sudo adduser itadmin`   // do not give full name and leave empty for all the fields

    `$ sudo adduser itadmin sudo`

    Now, username itadmin is created. Log into itadmin and do the rest.

2. Another important restriction is, the hostname of the nodes must be unique across the cluster; that is, the hostname should not be identical in the cluster. To change the hostname to ubuntu1 in your machine without restarting, use the following commands

    `$ sudo vi /etc/hostname`

    ```
        ubuntu1
    ```

    `$ sudo vi /etc/hosts`

    ```
        127.0.0.1    localhost
        127.0.1.1    ubuntu1
    ```

    `$ sudo sysctl kernel.hostname=ubuntu1`

    Do this in all nodes and create unique hostname as given in Table 6.1.

6.2.2   *INSTALLATION STEPS*

Step 1: Set the domain name in all the nodes in the cluster

Go to /etc/hosts to set DNS. After setting the domain name, you can use domain names instead of IP address. Comment 127.0.*.1 and enter IP, domain name, and hostname of each node in a line with tab space separator (refer Table 6.1).

```
$ sudo vi /etc/hosts
        #127.0.0.1        localhost
        #127.0.1.1        ubuntu1
        # IP              domain_name        host_name
        10.100.55.92        node1              ubuntu1
        10.100.55.93        node2              ubuntu2
        10.100.55.94        node3              ubuntu3
        10.100.55.95        node4              ubuntu4
        10.100.55.96        node5              ubuntu5
```

Step 2: Set passwordless communication

Create an SSH key in NN and send to all other nodes. Similarly, create an SSH key in RM and send it to all the other nodes. You need not create any key in slaves and SNN. In case, if you are hosting NN and RM in the same node, it is sufficient to do once.

Create key in node1 (NN) and send to all other nodes in the cluster

```
$ ssh-keygen or ssh-keygen -t rsa
$ cat ~/.ssh/id_rsa.pub >> ~/.ssh/authorized_keys
$ chmod 0700 ~/.ssh/authorized_keys
$ ssh-copy-id -i ~/.ssh/id_rsa.pub username@IP // to send keys to other
nodes
$ ssh-copy-id -i ~/.ssh/id_rsa.pub itadmin@node2
$ ssh-copy-id -i ~/.ssh/id_rsa.pub itadmin@node3
$ ssh-copy-id -i ~/.ssh/id_rsa.pub itadmin@node4
$ ssh-copy-id -i ~/.ssh/id_rsa.pub itadmin@node5
```

Try to remotely login from node1 to all other nodes in the cluster using ssh command without password. For example, from node1

```
$ ssh node2
$ exit
```

Similarly, create a ssh key in node2 (RM), send it to all the other nodes (1,3,4,5), and verify the passwordless connection.

Step 3 to Step 7 must be done in all the nodes in the cluster

Hadoop does not have a single, global location for configuration information. Instead, each node in the cluster has its configuration files, and it is up to the administrator to ensure that they are kept in sync at any point in time.

Step 3: Download Hadoop2.7.0 and move into hadoop folder

https://archive.apache.org/dist/hadoop/core/
$ wget https://archive.apache.org/dist/hadoop/core/hadoop-2.7.0/hadoop-2.7.0.tar.gz
$ tar -zxvf hadoop-2.7.0.tar.gz
$ sudo cp -r hadoop-2.7.0 /usr/local/hadoop

Step 4: Set path to Hadoop

$ vi .bashrc

```
export HADOOP_HOME=/usr/local/hadoop
export PATH=$PATH:$HADOOP_HOME/bin
export PATH=$PATH:$HADOOP_HOME/sbin
export HADOOP_HDFS_HOME=$HADOOP_HOME
export HADOOP_MAPRED_HOME=$HADOOP_HOME
export HADOOP_COMMON_HOME=$HADOOP_HOME
export YARN_HOME=$HADOOP_HOME
export HADOOP_CONF_DIR=$HADOOP_HOME/etc/hadoop
export HADOOP_COMMON_LIB_NATIVE_DIR=$HADOOP_HOME/lib/native
export HADOOP_OPTS="-Djava.library.path= $HADOOP_HOME/lib"
```

$ source .bashrc
$ $PATH

Step 5: Configure Java_Home, disable IPv6 in hadoop-env.sh

$ sudo vi /usr/local/hadoop/etc/hadoop/hadoop-env.sh

```
export JAVA_HOME=/usr/lib/jvm/java-8-openjdk-amd64
export HADOOP_OPTS=-Djava.net.preferIPv4Stack=true
```

$ yarn version

Step 6: Edit configuration files

To set up NN

$ sudo vi /usr/local/hadoop/etc/hadoop/core-site.xml

```
<configuration>
    <property>
        <name>fs.defaultFS</name>
        <value>hdfs://node1:10001</value>
    </property>
    <property>
        <name>hadoop.tmp.dir</name>
        <value>/usr/local/hadoop/tmp</value>
    </property>
</configuration>
```

To set up YARN

$ sudo vi /usr/local/hadoop/etc/hadoop/yarn-site.xml

```
<configuration>
    <property>
        <name>yarn.nodemanager.aux-services</name>
        <value>mapreduce_shuffle</value>
    </property>
    <property>
        <name>yarn.nodemanager.aux-
        services.mapreduce.shuffle.class</name>
        <value>org.apache.hadoop.mapred.
        ShuffleHandler</value>
    </property>
    <property>
        <name>yarn.resourcemanager.hostname</name>
        <value>ubuntu2</value>
    </property>
    <property>
        <name>yarn.resourcemanager.address</name>
        <value>node2:8030</value>
```

```
    </property>
    <property>
        <name>yarn.resourcemanager.resource-
        tracker.address</name>
        <value>node2:8031</value>
    </property>
    <property>
        <name>yarn.resourcemanager.scheduler.
        address</name><value>node2:8032</value>
    </property>
    <property>
        <name>yarn.resourcemanager.admin.address</
        name><value>node2:8033</value>
    </property>
    <property>
        <name>yarn.resourcemanager.webapp.address</
        name><value>node2:8088</value>
    </property>
    <property>
        <name>yarn.nodemanager.
        disk-health-checker.
        min-healthy-disks</name>
        <value>0.0</value>
    </property>
    <property>
        <name>yarn.nodemanager.
        disk-health-checker.
        max-disk-utilization-per-disk-percentage</
        name><value>98.5</value>
    </property>
</configuration>
```

To set up MR-related properties

```
$ sudo cp /usr/local/hadoop/etc/hadoop/mapred-site.xml.template /usr/
local/hadoop/etc/hadoop/mapred-site.xml
```

```
$ sudo vi /usr/local/hadoop/etc/hadoop/mapred-site.xml
    <configuration>
        <property>
            <name>mapreduce.framework.name</name>
            <value>yarn</value>
        </property>
        <property>
            <name>mapreduce.jobhistory.address</name>
            <value>node2:10020</value>
        </property>
        <property>
            <name>mapreduce.jobhistory.webapp.address</
            name><value>node2:19888</value>
        </property>
        <property>
            <name>mapreduce.jobhistory.
            intermediate-done-dir</name>
            <value>/usr/local/hadoop/tmp</value>
        </property>
        <property>
            <name>mapreduce.jobhistory.done-dir</name>
            <value>/usr/local/hadoop/tmp</value>
        </property>
    </configuration>
```

To set up NN, SNN, DN configuration (optional)

```
$ sudo vi /usr/local/hadoop/etc/hadoop/hdfs-site.xml
    <configuration>
        <property>
            <name>dfs.replication</name>
            <value>3</value>
        </property>
        <property>
            <name>dfs.http.address</name>
            <value>node1:50070</value>
        </property>
    </configuration>
```

To set up SNN: runs automatically with NN. But, it can be set as follows

```
$ sudo vi /usr/local/hadoop/etc/hadoop/masters
          node3
```

To set up slave (DN and NN)

```
$ sudo vi /usr/local/hadoop/etc/hadoop/slaves
              node3
              node4
              node5
```

Step 7: Change ownership and grant access to Hadoop services

```
$ sudo chown -R itadmin /usr/local/hadoop
$ sudo chmod -R 777 /usr/local/hadoop
```

Step 8: Create HDFS namespace (in node1)

Once all nodes are configured successfully, then format HDFS to create namespace.

```
$ hdfs namenode -format   // HDFS starts with root (/) file system.
```

Step 9: Start HDFS/YARN manually

In NN (node1)

```
$ start-dfs.sh
$ jps                     // run jps in slave nodes to see DN running
$ hdfs dfsadmin -report   // check storage status
$ stop-dfs.sh             // to stop HDFS
```

In RM (node2)

```
$ start-yarn.sh
$ mr-jobhistory-daemon.sh start historyserver
$ jps                 // run jps in slave nodes to see NM running
$ yarn node -list     // check YARN status, to check NN has started
$ stop-yarn.sh        // to stop YARN
```

Step 10: Run a simple wordcount program

```
$ hdfs dfs -mkdir /data
$ vi input.txt                       // enter some text
$ hdfs dfs -copyFromLocal input.txt /data
$ hdfs dfs -ls /data
```

```
$ yarn jar /usr/local/hadoop/share/hadoop/mapreduce/hadoop-
mapreduce-examples-2.7.0.jar wordcount /data/input.txt /output
$ hdfs dfs -cat /output/part-r-00000
```

To kill hanging tasks/applications

```
$ yarn application -list
$ yarn application -kill jobid
```

You can use Linux commands to kill the running container

```
$ jps                        // to see process id for running tasks
$ sudo kill -9 processid     // this will kill task
```

Access Hadoop services using web URL.

1. Web UI for Hadoop NN: http://node1:50070/
2. Web UI for application details: http://node2:8088/
3. Web UI for job history: http://node2:19888/
4. Web UI for NM: http://node3:8042/

Please refer to Appendix A and Ref. [18] for big data sources. Launching commands for HDFS and MR can be done from any node in the cluster as all nodes have the same configuration of master and slaves. Ultimately, you will get the same result. However, test with a small and large dataset to observe the running time that Hadoop takes. Hadoop takes more time for small files than processing with the local file system. Because there is a lot of intermediate steps involved, which takes a considerable amount of time. However, with the large dataset, you can observe Hadoop outperforming local file system processing. Therefore, upload GB size of data in the multi-node cluster and then launch MR wordcount job.

If the job produces huge output, then there will create many output files (each is an HDFS block) in the output directory. We have to open the file one by one to view the output. However, there is a way to view the entire output files as a single file.

```
$ hadoop fs -cat /output/part-*        // or bring them to local file system
$ hadoop fs -getmerge /output_dir_in_hdfs file_name_local
$ hadoop fs -getmerge /output result.txt
$ cat result.txt
```

As given in Table 6.2, you are not constrained to deploy HDFS and RM daemons in any different combinations depending upon the number of nodes available in the cluster.

TABLE 6.2 Different Choice of Multi-Node Implementation

Choices	node1	node2	node3	node4	node5
1	NN, SNN, RM, JHS, DN, NM	-	-	-	-
2	NN, SNN, RM, JHS	DN, NM	DN, NM	DN, NM	DN, NM
3	NN, SNN	RM	JHS	DN, NM	DN, NM
4	NN	SNN	RM	JHS	DN, NM
5	NN, DN, NM	SNN, DN, NM	RM, DN, NM	JHS, DN, NM	DN, NM

Table 6.3 shows the minimal set of properties to configure Hadoop deployment in short.

TABLE 6.3 Minimal Configuration for Hadoop Deployment

Daemon	Property	Local	Pseudo-distributed	Fully distributed
common	fs.defaultFS		hdfs://localhost/	hdfs://NN_IP:port#/
HDFS	dfs.replication	—	1	3 (default)
MR	mapreduce.framework.name	local	yarn	Yarn
YARN	yarn.resourceman-ager.hostnmae	—	localhost	RM_IP:port#
	yarn.nodemanager.aux-services	—	mapreduce_shuffle	mapreduce_shuffle

In order to experiment with huge data, you can download over 500 GB size of data from [21]. If it is not possible to download, then you can also generate a huge random text file with the size you wanted using the following command. To generate 5 GB text file, specify the size in bytes.

```
$ yarn jar /usr/local/hadoop/share/hadoop/mapreduce/hadoop-mapre-
duce-examples-2.7.0.jar teragen 5000000000 /terasort-input
```

6.3   HADOOP ADMINISTRATIVE ACTIVITIES

As a Hadoop administrator, one has several responsibilities to keep the cluster alive for application programers. Some of them are

- commissioning/decommissioning slaves.
- check for block corruption and copy block in and out across cluster.
- performance tuning to optimize the network, latency, etc.
- trouble-shooting jobs.
- upgrading to a newer version.

6.3.1   DECOMMISSIONING NODES

To increase the storage and computing capacity in a cluster, you will need to add or remove nodes from the cluster from time to time. Decommissioning is the process of removing slaves from Hadoop cluster while the cluster is up and running. Sometimes, it is necessary to decommission a node if it is failing more often or if its performance is noticeably slow. Some slaves may perform very poor, or you expect some slaves to die shortly. So, it is better to decommission such slaves from the cluster to preserve data blocks. NN replicates the blocks that were in the removed slaves to other DNs before respective DN is removed from the cluster.

1. Launch a multi-node cluster, as discussed in Section 6.2. Let us remove slave nodes 4 and 5.
2. Create a file called "exclude" in NN and RM. If NN and RM are running in the same machine, it is enough creating "exclude" file once. Include a list of domain names (or IPs) that you want to remove from the cluster.
   $ sudo vi /usr/local/hadoop/etc/hadoop/exclude

   ```
         node4
         node5
   ```
3. Edit hdfs-site.xml in NN to include the location of exclude file.
   $ sudo vi /usr/local/hadoop/etc/hadoop/hdfs-site.xml

   ```
      <configuration>
         <property>
            <name>dfs.hosts.exclude</name>
            <value>/usr/local/hadoop/etc/hadoop/
            exclude</value>
         </property>
   ```

```
    </configuration>
```
4. Edit yarn-site.xml in RM to include the location of exclude file.
    $ sudo vi /usr/local/hadoop/etc/hadoop/yarn-site.xml

```
    <configuration>
        <property>
            <name>yarn.resourcemanager.nodes.
            exclude-path</name>
            <value>/usr/local/hadoop/etc/hadoop/
            exclude</value>
        </property>
    </configuration>
```

5. Refresh RM to take effect of decommissioning NMs mentioned in yarn-site.xml. It means that RM will not create containers in decommissioned NMs.
    $ yarn rmadmin -refreshNodes
6. Refresh NN to take effect of decommissioning DNs mentioned in hdfs-site.xml. It means that NN will not direct any blocks to store in decommissioned DNs.
    $ hdfs dfsadmin -refreshNodes
7. Check WUI of NN and RM, to see the decrease in the count of nodes and resources.
8. If too many slave nodes are removed, it is better to balance the cluster. It will redistribute the blocks among available DNs.
    $ start-balancer.sh
9. To safely remove slaves from the cluster, stop DN+NM services in the respective slave and remove hostname/IP from slaves file in NN, and RM.
    $ hadoop-daemon.sh stop datanode
    $ yarn-daemon.sh stop nodemanager
    $ sudo vi /usr/local/hadoop/etc/hadoop/slaves

6.3.2   COMMISSIONING NODES

Adding new slave nodes to the existing Hadoop cluster to increase storage and computational capability is called commissioning nodes. It is a potential security risk to allow any machine to connect to the NN as a slave. Because new nodes may gain access to data that is not authorized to see. Such nodes are not under your control and may stop at any time, potentially causing data loss. Therefore, you have to be careful while adding slaves. Let us add the decommissioned nodes 4 and 5 back to the cluster.

1. If commissioning nodes are new nodes, complete the requirements given in Section 3.1 in node4 and node5. Similarly, configure Hadoop as given in Section 6.2 in the new nodes. Follow the multi-node installation Step1 to Step7. It is very important to set passwordless communication.

2. Add IP addresses of node4 and node5 in slaves file of both NN and RM for future cluster-wide operations. If both NN and RM run on the same machine, it is enough doing once.

    $ sudo vi /usr/local/hadoop/etc/hadoop/slaves

    ```
    node4
    node5
    ```

3. Remove domain names (node4, node5) from /usr/local/hadoop/etc/hadoop/exclude file if already included in it.

4. Create a file called "include" in NN and RM. If NN and RM are running in the same machine, it is enough creating "include" file once. Include a list of domain names (or IPs) that you want to add to the running cluster.

    $ sudo vi /usr/local/hadoop/etc/hadoop/include

    ```
    node4
    node5
    ```

5. Edit hdfs-site.xml in NN to include the location of include file.

    $ sudo vi /usr/local/hadoop/etc/hadoop/hdfs-site.xml

    ```
    <configuration>
        <property>
            <name>dfs.hosts</name>
            <value>/usr/local/hadoop/etc/hadoop/
            include</value>
        </property>
    </configuration>
    ```

6. Edit yarn-site.xml in RM to include the location of include file.

    $ sudo vi /usr/local/hadoop/etc/hadoop/yarn-site.xml

    ```
    <configuration>
        <property>
            <name>yarn.resourcemanager.nodes.
            include-path</name>
    ```

```
            <value>/usr/local/hadoop/etc/hadoop/
            include</value>
        </property>
    </configuration>
```

7. Start DN and NM in node4 and node5.
    $ hadoop-daemon.sh start datanode
    $ yarn-daemon.sh start nodemanager
    $ jps

8. Refresh RM to take effect of commissioning NMs mentioned in yarn-site.xml. It means that RM will create containers in commissioned NMs in the future.
    $ yarn rmadmin -refreshNodes

9. Refresh NN to take effect of commissioning DNs mentioned in hdfs-site.xml. It means that NN will direct blocks to store in commissioned DNs.
    $ hdfs dfsadmin -refreshNodes

10. Check WUI of NN and RM, to see the increase in the number of nodes and resources.

11. If too many slave nodes are commissioned, it is better to balance the cluster. It will redistribute the blocks among available DNs.
    $ start-balancer.sh

Note: while decommissioning nodes from Hadoop cluster, make sure slave's domain names are not in the /usr/local/hadoop/etc/hadoop/include file.

6.4   GATEWAY NODE (CLIENT NODE)

In production clusters, Hadoop nodes (NN, RM, slaves) should not be used as clients to submit job and upload big data as these activities cause performance down in Hadoop nodes. Therefore, we need to use one or more gateway nodes in the Hadoop cluster. Because, when you launch a job or copy data onto HDFS, a JVM is created with some amount of resources. Gateway nodes are also configured with Hadoop, but Hadoop services are not launched. Gateway nodes could be network mounted or a RAID configured to send or receive data between users and Hadoop cluster.

6.5   DEVELOPING MR JOBS USING ECLIPSE

So far, we have been executing a predefined wordcount job that is part of Hadoop distribution. Now, let us write a wordcount job manually using Eclipse and test in Eclipse itself without Hadoop environment.

6.5.1   WRITING MRV2 JOB USING ECLIPSE

Here, we are going to develop a wordcount MapReduce job using Eclipse, convert the job into a JAR, move into Hadoop environment, and launch the job. Good understanding of Java utility and collection packages help you write elegant MR jobs. The steps are

1.   Launch Hadoop single node/multi-node.
2.   Create input.txt file.
3.   Load input.txt onto HDFS.
4.   Make sure Hadoop environment and Eclipse use the same java version.
5.   Write MR wordcount job.
6.   Add Hadoop dependencies with your project (we usually define in pom.xml) in Eclipse.
7.   Create a jar file.
8.   Copy jar file to the gateway node.
9.   Run a simple MR wordcount job.
10. Review job output.

Moving big data onto HDFS
1.   I started Hadoop single node service.
2.   Create input.txt file.

```
$ vi input.txt
        hi how are you?
```

3.   Upload input.txt onto HDFS.

```
$ hadoop fs -mkdir /data
$ hadoop fs -copyFromLocal input.txt /data
$ hadoop fs -ls /data
```

MRv2 is backward compatible. So, MRv1 jobs also can be run on YARN. There is no much difference in the new MR API when compared to old MR API. The only change you can see is a programming construct of an MR job.

Writing MR jobs: I used Eclipse in Windows and moved the jar to Hadoop cluster.

4. Make sure Java version is the same in Eclipse and Hadoop environment.
5. Let us write an MR wordcount job in Eclipse and move into Hadoop cluster. Launch Eclipse in Windows (if you use Linux, it is fine) and follow the instructions given below.

File → new → java project → project name: MapReduce → Finish
Right click on project → new → class → name: WC → copy the following code [9]

```
import java.io.IOException;
import java.util.StringTokenizer;
import org.apache.hadoop.conf.Configuration;
import org.apache.hadoop.fs.Path;
import org.apache.hadoop.io.IntWritable;
import org.apache.hadoop.io.LongWritable;
import org.apache.hadoop.io.Text;
import org.apache.hadoop.mapreduce.Job;
import org.apache.hadoop.mapreduce.Mapper;
import org.apache.hadoop.mapreduce.Reducer;
import org.apache.hadoop.mapreduce.lib.input.
FileInputFormat;
import org.apache.hadoop.mapreduce.lib.output.
FileOutputFormat;public class WC{
    public static class UDFmapper extends
    Mapper<LongWritable, Text, Text, IntWritable>{
    private final static IntWritable one = new
    IntWritable(1);
    private Text word = new Text();
    public void map(LongWritable key, Text value,
    Context context)throws IOException,
    InterruptedException {
        String line = value.toString();
        StringTokenizer itr = new
        StringTokenizer(line);
    while (itr.hasMoreTokens()) {
        word.set(itr.nextToken());
        context.write(word, one);
```

```
            }
            }
        }
public static class UDFreducer extends
Reducer<Text,IntWritable, Text,IntWritable> {
    private IntWritable result = new IntWritable();
    public void reduce(Text key, Iterable<IntWritable>
    values, Context context) throws IOException,
    InterruptedException {
        int sum = 0;
        for (IntWritable val : values) {
            sum += val.get();
        }
        result.set(sum);
        context.write(key, result);
    }
}
public static void main(String[] args) throws Exception {
        Configuration conf = new Configuration();
        Job job = Job.getInstance(conf, "wordcount");
        job.setJarByClass(WC.class);
    job.setMapperClass(UDFmapper.class);
        job.setReducerClass(UDFreducer .class);
        job.setOutputKeyClass(Text.class);
        ob.setOutputValueClass(IntWritable.class);
        FileInputFormat.addInputPath(job,
        new Path(args[0]));
        FileOutputFormat.setOutputPath(job,
        new Path(args[1]));
        System.exit(job.waitForCompletion(true)? 0:1);
    }
}
```

In general, a job is a program written in some computer language. An MR job is a program written in java program. It is possible to write an MR job in other programming languages (C++, Python, Ruby, etc.) as well. There are three modules in a typical MR job: map function, reduce function, and driver (main) function. In map function, pre-processing logic is coded. In reduce function, core algorithms concepts are implemented. These map/reduce functions and MR properties are invoked from driver function. In general,

MR job execution begins by executing the driver function. Subsequently, functions responsible for each step of the MR execution sequence is invoked in order. Do not worry, MapReduce program constructs and art of writing MapReduce jobs are discussed in Section 6.7 to Section 6.9.

6. Add supporting jars to MR program from the following locations:

Right click on the project → select properties → java build path → libraries → add external jar

> hadoop-2.7.0\share\hadoop\common\→ selecthadoop-common-2.7.0.jar.
> hadoop-2.7.0\share\hadoop\hdfs\ → select hadoop-hdfs-2.7.0.jar.
> hadoop-2.7.0\share\hadoop\mapreduce\ → select hadoop-mapreduce-client-common-2.7.0.jar and hadoop-mapreduce-client-core-2.7.0.jar.
> hadoop-2.7.0\share\hadoop\yarn\ → select hadoop-yarn-client-2.7.0.jar, hadoop-yarn-common-2.7.0.jar, hadoop-yarn-api-2.7.0.
> hadoop-2.7.0\share\hadoop\tools\lib → select all jars from this file.

7. Create JAR

Right click on your project → export → Java → JAR → select project name "MapReduce" → specify a location → finish. Give "Job" as JAR name.

8. Move Job.jar into Hadoop VM or any one of the nodes in Hadoop cluster using WinSCP software.

9. To launch MR job from the node where Job.jar was moved, the syntax is

```
$ yarn jar jarname.jar driver_class_name /location_filename /output_
directory
$ yarn jar Job.jar WC /data/input.txt /output
```

Note: Every time you run, you have to give a different output directory name (which should not exist) in HDFS. Go to WUI to see job progress, statistics, and other HDFS information.

> NN_IP:50070   to see NN, DN, and block information.
> RM_IP:8088   to see running job information.
> RM_IP:19888   to see job history information.

10. The output directory is created in HDFS. Tow files are created in the output directory

```
$ hdfs dfs -ls /output
```
> `_SUCCESS`   // an empty file that indicates the job is done
> `part-r-00000`   // which contains job output

```
$ hdfs dfs -cat /output/part-r-00000
```

If there are many parts in the output directory, we need to view one by one. However, to view entire output parts as a single file, use the following command.

```
$ hdfs dfs -cat /output/part-*
$ hdfs dfs -getmerge /output_dir_in_hdfs file_name_local
$ hdfs dfs -getmerge /output result.txt
```

You can load data onto HDFS and launch a job from any node in the cluster as job client runs in every Hadoop node. Because we have set mapred.xml and core-site.xml in all nodes. Therefore, the job request is handed over to JT. Try running the MR program with a small amount of data first in local mode setup, so it will be easy for you to detect any errors and verify the logic of your program. Then go for launching in the multi-node cluster with huge data.

6.5.2   HADOOP DEVELOPMENT TOOLS (HDT)

It is painful to code MR job in Eclipse, convert to the jar, move into Hadoop node, and launch it. How would it be to launch MR jobs directly from Eclipse onto Hadoop environment? The HDT helps us to launch MR jobs directly from Eclipse onto Hadoop environment. Download hadoop-eclipse-plugin.jar for your Hadoop version. Assume we have a single node Hadoop 2.7.0 environment and use Eclipse on Ubuntu machine, which can be a node in the Hadoop cluster or any other computer (your PC/laptop) that can communicate with Hadoop nodes. IP of NN is 10.100.55.92 and username is itadmin.

1. I have Ubuntu 18 desktop version on my laptop. Remember, the username of your laptop must be the same as the common username used in the Hadoop cluster, and hostname also must be unique.
2. Install java and set path.
3. Download [19] Eclipse and install.
4. Download hadoop-eclipse connector "hadoop-eclipse-plugin-2.6.0.jar" [20].
5. Copy hadoop-eclipse-plugin-2.6.0.jar to Eclipse installed location → plugins folder.
6. Launch Eclipse → create a project and write wordcount job (refer Section 6.5.1).
7. In Eclipse, go to windows →perspective open perspective other → select MapReduce.

8. Now, you will see MR elephant symbol right top corner, as shown in Figure 6.3.
9. Select MR elephant logo on the right bottom corner to add Hadoop environment. You will get a window, as shown in Figure 6.4.
10. Edit only the following fields as specified in core-site.xml and yarn-site.xml.

> location name: hdfs
> MRv2 master host: 10.100.55.92   port: 8032 (default port number)
> DFS master: 10.100.55.92   port:10001 (given in core-site.xml)

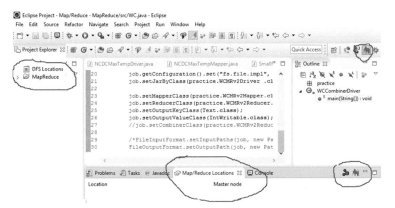

FIGURE 6.3   Configure HDT in Eclipse.

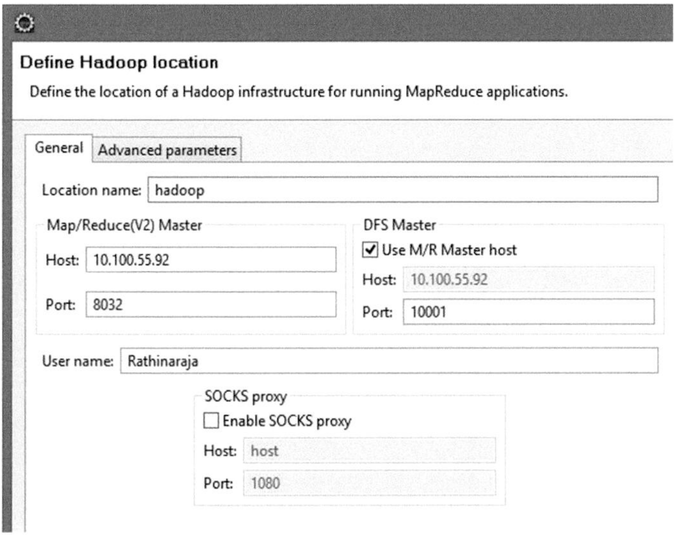

FIGURE 6.4   Enter NN and RM IP address and port number.

Note: If Hadoop is set up on a single node, give the same IP to both MRv2 and HDFS. In case of multi-node installation, you have to give RM IP address in MRv2 master host.

11. Go to advanced parameters and change only the following property
    – yarn.resourcemanager.address:10.100.55.92:8032 (it is the last property)
12. Now, in Eclipse project section, you can see HDFS files being displayed.
13. In MR driver function, the only change you have to do is to include the input file path (with HDFS URI) and the output directory path. Assume you have uploaded input.txt onto HDFS. Replace the following two lines in MR driver function.

```
FileInputFormat.addInputPath(job, new
Path(args[0]));
FileOutputFormat.setOutputPath(job, new
Path(args[1]));
```

with

```
FileInputFormat.addInputPath(job, new
Path("hdfs://10.100.55.92:10001/input.txt"));
FileOutputFormat.setOutputPath(job, new
Path("hdfs:// 10.100.55.92:10001/output"));
```

14. Click the run button in Eclipse to launch the job.
15. You can see the output file in Eclipse itself. Try refreshing or reconnecting HDFS if output files are not displayed.

6.5.3   RUN A SIMPLE MR JOB IN ECLIPSE ITSELF IN UBUNTU/ WINDOWS

Let us execute a simple MR wordcount job in Eclipse itself without any Hadoop environment. It is useful to test and debug the MR jobs easier and faster. If the MR job is executed in Eclipse, it is submitted to LocalJobRunner (no HDFS and MR daemons running to handle MR job). It is very similar to setting mapred.job.tracker is "local" for Hadoop standalone mode. MR framework itself applies the MR algorithm to execute a job. With Eclipse, you can launch an MR job on a small amount of input data to test the desired output and debug the jobs. The steps to run a simple MR wordcount job in Eclipse itself are:

Step 1: File → new → java project → project name: MapReduce → Finish
Step 2: Right-click on the project → new → class→ name:WC → write the following code [9]

```
import java.io.IOException;
import java.util.StringTokenizer;
import org.apache.hadoop.conf.Configuration;
import org.apache.hadoop.fs.Path;
import org.apache.hadoop.io.IntWritable;
import org.apache.hadoop.io.LongWritable;
import org.apache.hadoop.io.Text;
import org.apache.hadoop.mapreduce.Job;
import org.apache.hadoop.mapreduce.Mapper;
import org.apache.hadoop.mapreduce.Reducer;
import org.apache.hadoop.mapreduce.lib.input.File Input
    Format;
import    org.apache.hadoop.mapreduce.lib.output.File    Output
    Format;
public class WC{
    public static class UDFmapper extends Mapper<LongWritable,
    Text, Text, IntWritable>{
       private final static IntWritable one = new
       IntWritable(1);
       private Text word = new Text();
       public void map(LongWritable key, Text value, Context
       context)throws IOException, InterruptedException {
          String line = value.toString();
          StringTokenizer itr = new StringTokenizer(line);
       while (itr.hasMoreTokens()) {
          word.set(itr.nextToken());
          context.write(word, one);
       }
       }
    }
    public static class UDFreducer extends
    Reducer<Text,IntWritable, Text,IntWritable> {
       private IntWritable result = new IntWritable();
       public void reduce(Text key, Iterable<IntWritable>
       values, Context context) throws IOException, Interrupt-
       edException {
          int sum = 0;
          for (IntWritable val : values) {
              sum += val.get();
          }
          result.set(sum);
```

```
        context.write(key, result);
    }
}
public static void main(String[] args) throws
Exception {
    if(args.length <2) {
        System.err.println("Usage:<inputpath>
        <outputpath>");System.exit(-1);
    }
    Configuration conf = new Configuration();
    Job job = Job.getInstance(conf, "wordcount");
    job.setJarByClass(WC.class);
  job.setMapperClass(UDFmapper.class);
    job.setReducerClass(UDFreducer .class);
    job.setOutputKeyClass(Text.class);
    job.setOutputValueClass(IntWritable.class);
    FileInputFormat.addInputPath
    (job, new Path(args[0]));
    FileOutputFormat.setOutputPath(job, new
    Path(args[1]));
    System.exit(job.waitForCompletion(true)? 0:1);
    }
}
```

Step 3: Add supporting jars to MR program from the following locations:

Right-click on the project→ select properties→ java build path→ libraries→ add external jar→

Case 1: While using Eclipse on Ubuntu

If you use Hadoop 2.7.0, add the following library jars into Eclipse.
  hadoop-2.7.0\share\hadoop\common\→ select hadoop-common-2.7.0.jar.
  hadoop-2.7.0\share\hadoop\hdfs\ → select hadoop-hdfs-2.7.0.jar.
  hadoop-2.7.0\share\hadoop\mapreduce\ → select hadoop-mapreduce-client-common-2.7.0.jar and hadoop-mapreduce-client-core-2.7.0.jar.
  hadoop-2.7.0\share\hadoop\yarn\ → select hadoop-yarn-client-2.7.0.jar, hadoop-yarn-common-2.7.0.jar, hadoop-yarn-api-2.7.0.
  hadoop-2.7.0\share\hadoop\tools\lib → select all jars from this file.

If you use Hadoop 1.2.1, add the following library jars into Eclipse.
  hadoop-1.2.1\lib → select all jars.

hadoop-1.2.1\ → select hadoop-client-1.2.1.jar and hadoop-core-1.2.1.jar

Case 2: While using Eclipse on Windows

When I added Hadoop 2.7.0 libraries, it threw some unknown exception and did not work. Therefore, I used Hadoop 1.2.1 library jars for both Hadoop 2.7.0 and Hadoop 1.2.1 to run MR job in Eclipse itself. However, if you want to convert Hadoop 2.7.0 MR jobs into a jar and launch onto real Hadoop cluster, add appropriate Hadoop version library jars.

Step 4: Passing input and output locations as arguments

When you execute MR job in Eclipse itself, you have to pass input and output locations as arguments to the job (very similar to launching jobs in the command line). There are two ways to achieve this.

Way1: Go to run → run configurations → double click java applications → select WC → select arguments → enter input file location and output directory location. For instance, I provide files location in Windows as below.

"D:/Eclipse Project/MapReduce/input.txt" "D:/Eclipse Project/ MapReduce/output"

Way2: You can type the input file and output directory path in the driver program itself.

```
FileInputFormat.setInputPaths(job,new Path
    ("D:/Eclipse Project/MapReduce/input.txt"));
FileOutputFormat.setOutputPath(job,new Path
    ("D:/Eclipse Project/MapReduce/output"));
```

But, you have to remove the following section from driver function

```
if(args.length <2) {
    System.err.println("Usage:<inputpath>
    <outputpath>"); System.exit(-1);
    }
```

Step 5: Run and see the output location.

This procedure just works fine in Ubuntu. However, when you run Hadoop in Eclipse on Windows, you have to create one more java program as given below.

Right click on MapReduce project → new → class → name: WinLocalFile-System → copy the following code [10].

```
import java.io.IOException;
import org.apache.hadoop.fs.LocalFileSystem;
import org.apache.hadoop.fs.Path;
import org.apache.hadoop.fs.permission.FsPermission;
```

```
public class WinLocalFileSystem extends
LocalFileSystem {
    public WinLocalFileSystem() {
        super();
        System.err.println("Patch for HADOOP-7682:
        "+"Instantiating workaround file system");
    }
    @Override
    public boolean mkdirs(Path path, FsPermission permis-
    sion)throws IOException {
            boolean result=super.mkdirs(path);
            this.setPermission(path,permission);
            return result;
    }
}
@Override
public void setPermission(Path path, FsPermission permis-
sion) throws IOException {
    try {
            super.setPermission(path,permission);
    }
    catch (IOException e) {
        System.err.println("Patch for HADOOP-7682: "+
        "Ignoring IOException setting persmission for path
        \""+path+ "\": "+e.getMessage());
            }
    }
}
```

Then, add the following line in main() of WC.java right after job object creation.

```
Job job = Job.getInstance(conf, "wordcount");
job.getConfiguration().set("fs.file.impl",
"WinLocalFileSystem");
```

Now, run the job. It just works and check the output file location.

Note: Make sure JDK version in Eclipse and Hadoop environment are the same before converting the job into a jar and launch onto Hadoop environment.

If you want to convert MRv1 job into a jar and run in Hadoop 1.2.1, add Hadoop 1.2.1 libraries. Similarly, add Hadoop 2.7.0 libraries for MRv2 jobs to run on Hadoop 2.7.0 before converting into the jar. Do not add different Hadoop version libraries together, which will cause an exception.

Importantly, do not forget to comment the following line in drive code before converting the MR job into the jar as it is used only when the MR job is executed in Eclipse on Windows.

```
// job.getConfiguration().set("fs.file.impl",
"WinLocalFileSystem");
```

6.6   HDFSV2 AND YARN COMMANDS

The file system shell is an interface between user and HDFS.

URI is used to locate resources like files. All the file system shell commands take URIs as arguments. Example:

- HDFS: hdfs://NN_IP:port#/file_location
- HAR: Hadoop archive har:///location/file.har
- Local file system: file:///location/file_name

Every command returns exit code either 0 for success or -1 for error.

```
$ hdfs dfs                          // dfs indicates we are working with HDFS
$ hdfs dfs -help                    // lists short summary of commands in dfs
$ hdfs dfs -help commandName // displays summary of specific command
$ hdfs dfs -help put
```

1. Create user directory and list meta-data (username of my machine is itadmin). In the HDFSv2, we have to create home directory (/user/itadmin) manually as soon as HDFS is up and running unlike in HDFSv1.

```
$ hdfs dfs -ls                             // no access to user home
$ hdfs dfs -mkdir /user
$ hdfs dfs -mkdir /user/itadmin
$ hdfs dfs -ls // now you can access user home
$ hdfs dfs -chown itadmin:itadmin /user/itadmin
                                           // providing ownership
$ hdfs dfsadmin -setSpaceQuota 1t /user/itadmin
                                           // to set 1TB limit on given user dir
$ hdfs dfs -mkdir dir
$ hdfs dfs -ls
    dir
```

```
$ hdfs dfs -mkdir /dir1          // to create directory in HDFS root
$ hdfs dfs -ls /                 // displays  meta-data  of  hadoop  root
dir
    /user
    /dir1
$ hdfs dfs -ls hdfs://NN_IP:10001/ // in case of multi node, use URI.
$ hdfs dfs -ls -R /              // shows hidden files and sub directories.
$ hdfs dfs -ls file:///          // lists out local file system root
```

You can create an empty file on HDFS but cannot edit any files in HDFS. You must create in local file system and copy to HDFS. Therefore, vi, gedit will not work with HDFS.

```
$ hadoop dfs -touchz /finput.txt     // file size is zero
$ vi input.txt                       // create a file in local file system
    Hi how are you?
$ hadoop dfs -ls /dir1
```

2. To copy/move files from local file system to HDFS

```
$ hdfs dfs -put local_source /hdfs_destination
$ hdfs dfs -copyFromLocal local_source /hdfs_destination
$ hadoop dfs -copyFromLocal input.txt /dir1
$ hdfs dfs -moveFromLocal local_source /hdfs_destination
$ hadoop dfs -moveFromLocal input.txt /dir1
```

   If source is file, destination can be a file or directory
   If source is directory, destination must be a directory

put vs copyFromLocal: put can copy more than one input file and read directly from stdin.

3. To copy/move files from HDFS to local file system

```
$ hdfs dfs -get /hdfs_source local_destination/filename
$ hdfs dfs -copyToLocal /hdfs_source local_destination
$ hdfs dfs -copyToLocal /dir1/input.txt ~/
$ hdfs dfs -moveToLocal /hdfs_source local_destination
$ hdfs dfs -moveToLocal /dir1/input.txt ~/
```

4. To copy/move a file from one location to another location in HDFS itself

```
$ hdfs dfs -cp /hdfs_source /hdfs_destination
$ hdfs dfs -cp /dir1/input.txt /
$ hdfs dfs -mv /hdfs_source /hdfs_destination
$ hdfs dfs -mv /dir1/input.txt /
```

5. To display a file from HDFS

```
$ hdfs dfs -cat file:///filename
$ hdfs dfs -cat /dir1/input.txt                    location of file in HDFS
$ hdfs dfs -cat hdfs://NN_IP:10001/dir1/input.txt
```

// URI of file in HDFS

6. To delete a file, directory from HDFS

```
$ hdfs dfs -rm /filename
$ hdfs dfs -rmr /directory
```

7. appendToFile – many files are concatenated while uploading onto HDFS into one file.

```
$ vi file1.txt
```

```
    hi how are you
```

```
$ vi file2.txt
```

```
    how do you do
```

```
$ hdfs dfs -appendToFile file1.txt file2.txt /new_filename
$ hdfs dfs -appendToFile file1.txt file2.txt /dir1/file3.txt
$ hdfs dfs -cat /dir1/file3.txt
```

```
    hi how are you
    how do you do
```

8. Checksum: Returns the checksum information of a file.

```
$ hdfs dfs -checksum hdfs://NN_IP:10001/dir1/input.txt
$ hdfs dfs -checksum /dir1/input.txt
```

```
    MD5-of-0MD5-of-
    512CRC32C000002000000000000004447e349865436d7e2
```

9. chmod

```
$ hdfs dfs -chmod 777 /filename
$ hdfs dfs -chmod 777 /dir1/input.txt
```

10. chown

    $ hdfs dfs -chown username /filename

    $ hdfs dfs -chown itadmin /dir1/input.txt

11. To display the size of directories/files in bytes.

    $ hdfs dfs -du /

    ```
    33      /dir1

    0       /user
    ```

    $ hdfs dfs -df /

    Filesystem Size Used Available Use% hdfs://node1:10001
    237560135680 675840 93917169664 0%

12. To get file path

    $ hdfs dfs -find / -name input.txt -print

    ```
    /dir1/input.txt
    ```

13. To display last 1 KB from a file to stdout (console).

    $ hdfs dfs -tail /dir1/input.txt

14. To display top or bottom "n" number of lines from a file in HDFS (add more lines in input.txt file).

    $ hdfs dfs -cat /dir1/input.txt | head -n 5   // displays top five lines

    $ hdfs dfs -cat /dir1/input.txt | tail -n 5   // displays bottom five lines

15. To count the number of files and the number of lines in a file in HDFS.

    $ hdfs dfs -count -q -h /

    $ hdfs dfs -ls / | wc -l   // to display the number of files in HDFS/.

    $ hdfs dfs -cat /input.txt | wc -l // to display number of lines/records in a file

16. Text takes a source file and outputs the file in text format.

    $ hdfs dfs -text /file_name

17. To list top 5 biggest/smallest size files in HDFS.

    $ hdfs dfs -du / |sort -g -r| head -n 5

    $ hdfs dfs -du / |sort -g -r| tail -n 5

18. Other file system operations

    $ hdfs dfsadmin -report   // displays BR

$ hdfs fsck /hdfs_directory -files -blocks -locations

$ hdfs dfs -setrep number /filename   // replication by default 3

19. YARN commands

    $ yarn   // shows sub commands

    $ yarn application

    $ yarn application -list   // lists running applications

    $ yarn application -kill <application_id>

    $ yarn application -list all   // returns history of jobs

20. Working with jars

    $ yarn jar /usr/local/hadoop/hadoop-examples-1.2.1.jar

                                // displays the jobs in jar

    $ jar tvf name.jar   // to list what is inside jar

21. To transfer files from one node to another node, make sure the target location has write permission of its current user.

    $ sudo chown -R itadmin /usr/local/hadoop

                            // at location of target machine

    $ sudo chmod -R 777 /usr/local/hadoop

                            // at location of target machine

    $ scp -r directory username@IP:~/   // this is from our machine

22. To execute commands in a remote node without logging into remote node

    $ ssh username@IP "ps -ef | grep -i namenode"

23. To find NN machine in the cluster

    $ hdfs getconf -namenodes   // finds the NN's hostname from fs.defaultFS

    $ hdfs getconf -secondarynamenodes

Check the following link for more commands
https://hadoop.apache.org/docs/r2.7.3/hadoop-yarn/hadoop-yarn-site/Yarn-Commands.html

YARN distributed shell

The YARN comes out of the box with two applications: MRv2 and distributed shell. Distributed shell is an example of a non-MR application built on top of YARN. Let us create a container in a Hadoop node to run the shell command (uptime command).

$ yarn org.apache.hadoop.yarn.applications.distributedshell.Client
-jar /usr/local/hadoop/ share/hadoop/yarn/hadoop-yarn-applications-
distributedshell-2.7.0.jar -container_memory 350 -master_memory 350
-shell_command uptime

Distcp

Distcp (distributed copy) is a tool used for large inter/intra-cluster copying. It runs as an MR job (only map tasks, no reduce tasks) to achieve distribution, error handling, recovery, and reporting. It is mainly used when you want to copy data among Hadoop cluster in HDFS federation. It copies data blocks in parallel across clusters.

$ yarn distcp hdfs://NN1_IP/source_file hdfs://NN2_IP/desti_dir

6.7   PROPERTIES IN HADOOP 1.X AND HADOOP 2.X

In nearly all cases, an MR job will either encounter a bottleneck reading data from disk or the network (IO-bound job) or processing data (CPU-bound). Examples of an IO-bound job is sorting, which requires very little processing (simple comparisons) and much reading and writing to disk. An example of a CPU-bound job is machine learning algorithms, in which input data is processed in a very complex way to extract knowledge. Therefore, we need to tune software and hardware-level properties to gain more performance for MR jobs.

Configuration class (**org.apache.hadoop.conf** package) contains a collection of configuration properties. A property is represented by a pair of name-value. Property name is String type, and property value can be of boolean, int, long, float, String, Class, etc.

```
<configuration>
    <property>
        <name> </name>
        <value> </value>
    </property>
</configuration>
```

There are two types of properties in general: system (JVM) property, job configuration property. core-site.xml, yarn-site.xml, mapred-site.xml, hdfs-site.xml files in /usr/local/ hadoop/etc/hadoop location are used to configure Hadoop daemons and control MR job execution sequence. However, some default properties are already included in core-default. xml, yarn-default.xml, mapred-default.xml, and hdfs-default.xml files which are available at /usr/ local/ hadoop/share/doc/hadoop/

- hadoop-yarn/hadoop-yarn-common/yarn-default.xml
- hadoop-project-dist/hadoop-common/core-default.xml
- hadoop-project-dist/hadoop-hdfs/hdfs-default.xml
- hadoop-mapreduce-client/hadoop-mapreduce-client-core/mapred-default.xml

These properties can be overridden at run time via command line while launching job or can be hard-coded in job driver function. JVM properties can be retrieved using java.lang.System class. Job configuration properties are retrieved using Configuration object. Please note that some properties of map/reduce task cannot be overridden at run time. Therefore, it must be set before Hadoop services start.

The property naming convention of Hadoop 1.x is not followed in Hadoop 2.x. Because porting MRv2 with YARN has brought up some major changes in naming convention of MRv2 properties. Let us see some important MR and container-level new properties.

6.7.1 MR-LEVEL PROPERTIES IN MRV2

mapreduce.framework.name: determines the mode of MR deployment. The default value is local. Possible options are:

local – means that LocalJobRunner is used (MR execution is done in single JVM).

classic – means that MR job will be launched on MRv1 cluster. In this case, mapreduce. jobtracker.address property is used to specify JT.

yarn – runs MR jobs on YARN, which can be either pseudo-distributed mode or multi-node cluster.

mapreduce.job.ubertask.enable: If jobs are small, they can run in MRApp-Master JVM itself to avoid the overhead of spawning and scheduling task containers. The default value is false.

mapreduce.shuffle.max.connections: Denotes the maximum allowed connections for shuffle. The default value is 0, which indicates that no limit on the number of connections.

yarn.resourcemanager.am.max-attempts: A maximum number of attempts for MRAppMaster. The default value is 2. This can be overridden by individual jobs and cannot exceed 2.

yarn.resourcemanager.recovery.enabled: enables RM to recover state after restarting. The default value is false. If true, yarn.resourcemanager.store. class must be set. There is a possibility to set zookeeper-based mechanism to store RM state (org.apache.hadoop.yarn. server.resourcemanager.recovery. ZKRMStateStore).

yarn.resourcemanager.store.class: writes RM state into a file system for recovery purposes.

6.7.2   CONTAINER-LEVEL PROPERTIES IN MRV2

Related to map/reduce tasks

mapredue.map.memory.mb: amount of memory given to map task in MBs, by default 1024.

mapreduce.reduce.memory.mb: amount of memory given to reduce task, by default 1024.

mapreduce.map.cpu.vcores: number of logical cores allocated to map task, by default 1.

mapreduce.reduce.cpu.vcores: number of logical cores allocated to reduce task, by default 1.

mapreduce.child.java.opts: heap space in JVM given for MR tasks, by default Xmx200m.

6.7.3   MRV1 PROPERTIES THAT ARE NO LONGER EFFECTIVE IN MRV2

mapred.job.tracker: No JT in YARN. Most of the JT functions are done by MRAppMaster.

mapred.local.dir: to store intermediate data for MR jobs.

mapred.tasktracker.map/reduce.tasks.maximum: the number of map/reduce slots in a TT.

tasktracker.http.threads: map outputs are fetched by ShuffleHandler in MRv2 and by default configured with no cap in the number of open connections (new name is mapreduce.shuffle. max.connections) io.sort.record.percent: this property is used to specify the buffer size (using io.sort.mb) for map side sort. MRv2 is smarter to fill io.sort.mb.

6.7.4 DEPRECATED PROPERTIES IN MRV2

Most of the MRv1 properties are deprecated in MRv2. Because MRv2 properties are well organized to suit the YARN environment. However, Hadoop v2 supports both old and new properties. Let us see some of the important properties that are renamed from Hadoop 1.x to Hadoop 2.x to adapt YARN architecture as given in Table 6.4.

TABLE 6.4   Hadoop 1.x Vs Hadoop 2.x Properties [11]

Hadoop 1.x	Hadoop 2.x
dfs.block.size	dfs.blocksize
dfs.data.dir	dfs.datanode.data.dir
dfs.name.dir	dfs.namenode.name.dir
dfs.name.edits.dir	dfs.namenode.edits.dir
dfs.permissions	dfs.permissions.enabled
fs.checkpoint.dir	dfs.namenode.checkpoint.dir
fs.checkpoint.period	dfs.namenode.checkpoint.period
fs.default.name	fs.defaultFS
io.sort.factor	mapreduce.task.io.sort.factor
io.sort.mb	mapreduce.task.io.sort.mb
io.sort.spill.percent	mapreduce.map.sort.spill.percent
job.local.dir	mapreduce.job.local.dir
local.cache.size	mapreduce.tasktracker.cache.local.size
mapred.compress.map.output	mapreduce.map.output.compress
mapred.data.field.separator	mapreduce.fieldsel.data.field.separator
mapred.heartbeats.in.second	mapreduce.jobtracker.heartbeats.in.second
mapred.hosts.exclude	mapreduce.jobtracker.hosts.exclude.filename
mapred.hosts	mapreduce.jobtracker.hosts.filename
mapred.jar	mapreduce.job.jar
mapred.job.id	mapreduce.job.id
mapred.job.map.memory.mb	mapreduce.map.memory.mb
mapred.job.name	mapreduce.job.name
mapred.job.priority	mapreduce.job.priority
mapred.job.queue.name	mapreduce.job.queuename
mapred.job.reduce.input.buffer.percent	mapreduce.reduce.input.buffer.percent
mapred.job.reduce.memory.mb	mapreduce.reduce.memory.mb
mapred.job.shuffle.input.buffer.percent	mapreduce.reduce.shuffle.input.buffer.percent

TABLE 6.4   *(Continued)*

Hadoop 1.x	Hadoop 2.x
mapred.job.shuffle.merge.percent	mapreduce.reduce.shuffle.merge.percent
mapred.job.tracker.handler.count	mapreduce.jobtracker.handler.count
mapred.job.tracker.http.address	mapreduce.jobtracker.http.address
mapred.job.tracker	mapreduce.jobtracker.address
mapred.local.dir	mapreduce.cluster.local.dir
mapred.map.child.java.opts	mapreduce.map.java.opts
mapred.map.max.attempts	mapreduce.map.maxattempts
mapred.map.output.compression.codec	mapreduce.map.output.compress.codec
mapred.mapoutput.key.class	mapreduce.map.output.key.class
mapred.mapoutput.value.class	mapreduce.map.output.value.class
mapred.map.tasks.speculative. execution	mapreduce.map.speculative
mapred.max.map.failures.percent	mapreduce.map.failures.maxpercent
mapred.max.reduce.failures.percent	mapreduce.reduce.failures.maxpercent
mapred.max.split.size	mapreduce.input.fileinputformat.split.maxsize
mapred.output.compression.codec	mapreduce.output.fileoutputformat.compress.codec
mapred.output.compression.type	mapreduce.output.fileoutputformat.compress.type
mapred.output.compress	mapreduce.output.fileoutputformat.compress
mapred.output.dir	mapreduce.output.fileoutputformat.outputdir
mapred.reduce.child.java.opts	mapreduce.reduce.java.opts
mapred.reduce.max.attempts	mapreduce.reduce.maxattempts
mapred.reduce.parallel.copies	mapreduce.reduce.shuffle.parallelcopies
mapred.reduce.slowstart.completed. maps	mapreduce.job.reduce.slowstart.completedmaps
mapred.reduce.tasks	mapreduce.job.reduces
mapred.reduce.tasks.speculative. execution	mapreduce.reduce.speculative
mapred.skip.map.max.skip.records	mapreduce.map.skip.maxrecords
mapreduce.combine.class	mapreduce.job.combine.class
mapreduce.inputformat.class	mapreduce.job.inputformat.class
mapreduce.outputformat.class	mapreduce.job.outputformat.class
mapreduce.partitioner.class	mapreduce.job.partitioner.class
mapreduce.reduce.class	mapreduce.job.reduce.class
mapred.working.dir	mapreduce.job.working.dir
mapred.work.output.dir	mapreduce.task.output.dir

Reprinted from Ref. [11].

6.7.5 TOOL AND GENERIC OPTIONS PARSER

You can pass run-time input to the job via command line arguments, but you have to hard code to receive arguments in job's driver function. If the number of input arguments is dynamic, you cannot hard-code to receive the arguments every time. Therefore, Hadoop provides some helper classes and interfaces such as GenericOptionsParser, Tool, etc.

- GenericOptionsParser is a class that takes command line arguments and assigns to the Configuration object along with job properties.
- The tool is an interface that runs job with ToolRunner class. This provides a facility to pass generic options via CLI during run time and uses GenericOptionsParser internally.

There are different generic options (-D, -conf, -fs, -jt, -files, -libjars, -archives) to pass different types of input via command line while launching MR jobs. These generic options are obtained by getRemainingArgs(). The functionality of GenericOptionsParser is to segregate the generic options from command line arguments like input, output. Some of these are mentioned below:

-D <property=value>   To pass job-level property one by one.
-conf <configuration file>   To pass more than one property via .xml file.
-fs <local or NN_IP:port#>   To specify NN.
-jt <local or JT:port#>   To specify JT.
-files <comma separated list of files>
                             To specify a list of files to copy into HDFS.
 -libjars <comma separated list of jars>
                 To specify a list of jar files to include in the job classpath.
 -archives < list of archives>
                  To specify a list of archives (like jar files) to pass to NMs.

Let us see some examples using this facility.

1. To display the content of a namespace if there is more than one NNs.

   $ hdfs dfs -D fs.defaultFS=NN_IP:port# -ls /
   $ hdfs dfs -D fs.defaultFS=node1:10001 -ls /   **or**
   $ hdfs dfs -fs NN_IP:port# -ls /
   $ hdfs dfs -fs node1:10001 -ls /

2. If there is more than one JT in MRv1, you can specify which JT job should be submitted.

$ hadoop jar job.jar classname -jt JT_IP:port# /input /output
$ hadoop jar job.jar WC -jt node2:10002 /input /output

3. To pass supporting jars and archived files along with MR job

    $ yarn jar job.jar classname -libjars testlib.jar -archives test.tgz -files file.
    txt /input /output

4. To disable speculative execution

    $ yarn jar job.jar classname -D mapreduce.map.speculative=false /
    input /output

5. To specify specific RM address if there are more than RM

    $ yarn jar job.jar classname -D yarn.resourcemanager.address=IP:port#
    /input /output

Properties passed at run time overrides the properties already defined in Hadoop configuration files. However, if a property is defined as final in the configuration file, it cannot be overridden at run time. Similarly, configurations such as defining NM, JVM cannot be overridden at run time. JVM properties are looked up using java.lang.System class while Hadoop properties are looked up using Configuration object with GenericOptionsParser.

6.7.6 SETTING CONFIGURATION PROPERTIES

YARN, MR properties can be set statically/dynamically overridden at run time. This flexibility should be carefully used as it can result in a catastrophic outcome. There are many different ways to achieve this. Let us see one by one.

1. XML configuration files

    Before starting Hadoop services, you can configure properties in *-site.
    xml files located in `/usr/local/hadoop/etc/hadoop/`. Once properties are configured in these files, this is common for all the jobs and permanent until you reconfigure it. Modifying properties in these files require restarting services. We have discussed enough to set properties in these configuration files during Hadoop installation.

2. hadoop-env.sh and yarn-env.sh

    System and Hadoop service-related configurations can be done using hadoop-env.sh file also. This is also a permanent configuration and is not job-specific. Example:

```
$ vi /usr/local/hadoop/etc/hadoop/hadoop-env.sh
    export HADOOP_NAMENODE_OPTS="-XX:+UseParallelGC"
```

3. Overriding properties at run time

    We discussed enough this in the last section. Job-specific properties can be passed at run time. It overrides the properties that are already in configuration files. -D is used to pass job-specific parameters. Example:

    To specify the combiner class while launching job
    ```
    $ yarn jar name.jar -D mapreduce.job.combine.class=WCReducer /
    input /output
    ```

    To set RF for files while uploading onto HDFS
    ```
    $ hdfs dfs -D dfs.replication=2 -put input_file /dir
    ```

    To replicate a file already existing in HDFS
    ```
    $ hdfs dfs -setrep 5 /file_name
    ```

    To customize the RF for output file of a job
    ```
    $ yarn jar name.jar classname -D dfs.replication=10 /input /output
    ```

However, if a property is specified as final in configuration files, it cannot be overridden at run time. Example: RF specified in hdfs-site.xml cannot be overridden at run time.

```
<configuration>
    <property>
        <name>dfs.replication</name>
        <value>2</value>
        <final>true</final>
    </property>
</configuration>
$ hdfs dfs -D dfs.replication=2 -put inputfile /dir // exception is thrown
```

4. conf.xml: overriding parameters at run-time in command line can pass only one property. To override a set of properties at run time, we can use conf.xml file.

    ```
    $ vi conf.xml
        <?xml version="1.0">
        <configuration>
            <property>
                <name>dfs.replication</name>
                <value>2</value>
            </property>
    ```

```
        <property>
            <name>dfs.namenode.name.dir</name>
            <value>/usr/local/Hadoop/tmp/dfs/name</
            value>
        </property>
    </configuration>
```
$ yarn jar name.jar classname -conf conf.xml /input /output

5. Job driver function

    We can set the properties to be job-specific in driver program. Example:
    in MRv2 job

```
Configuration conf = new Configuration();
Job job = Job.getInstance(conf, "wordcount");
job.getConfiguration().set("fs.defaultFS", "hdfs://
NN_IP:10001");
job.getConfiguration().set("mapreduce.framework.
name", "yarn");
```

6.8   HADOOP APIS

Application Programming Interface (API) is a bunch of predefined classes
that you can use readily at any time. Example: to print something on screen
using java program, rather than coding yourself it is good to use System.
out.println(). Hadoop has a plenty of APIs in different packages https://
hadoop.apache.org/docs/r2.7.0/api/ to support MR execution. You need to
understand Hadoop API for writing custom MR job and to override MR
execution sequence such as shuffle, sort, and group. org.apache.hadoop is the
main package, under which there are three major sub-categories: common,
mapreduce, yarn.

*   **Common:** It includes packages for the file system (HDFS, S3, azure),
    utility, io, configurations, security that support both MR and non-MR
    applications.

    org.apache.hadoop.fs   // to perform file-related operations on HDFS.
    org.apache.hadoop.io   // to read write from HDFS and other file systems.
    org.apache.hadoop.util   // contains tool interface
    org.apache.hadoop.fs.s3   // for AWS S3 storage
    org.apache.hadoop.fs.azure   // for Azure blob storage

- **Mapreduce:** This package provides APIs for writing MR jobs. Sub packages of mapreduce are lib, jobcontrol, etc.

  org.apache.hadoop.mapred   // old API (MRv1)
  org.apache.hadoop.mapreduce   // new API (MRv2)

  You can distinguish new and old version API with font style: *Mapper* in italic is old API, Mapper in normal letter is new API. MRv2 is backward compatible to MRv1, so MRv1 jobs also can be executed on YARN. However, do not mix both old and new API in the same job. It leads to confusion for MR job executor to pick API from right package at run time.

- **Yarn:** This package provides support for MR and non-MR application execution. Sub packages of yarn are client, conf, logaggregation, util, etc. You can see all these packages as jars in your downloaded Hadoop file.

  $ ls hadoop-2.7.0/share/hadoop/

  ```
  common httpfs hdfs kms mapreduce tools yarn
  ```

Minimum jars to be added for MRv2 job to run in pseudo-distributed or cluster mode are:

hadoop-2.7.0\share\hadoop\common\lib → select all jars from this file.
hadoop-2.7.0\share\hadoop\common\→ selecthadoop-common-2.7.0.jar
hadoop-2.7.0\share\hadoop\hdfs\ → select hadoop-hdfs-2.7.0.jar
hadoop-2.7.0\share\hadoop\mapreduce\ → select hadoop-mapreduce-client-common-2.7.0.jar and hadoop-mapreduce-client-core-2.7.0.jar
hadoop-2.7.0\share\hadoop\yarn\ → select hadoop-yarn-client-2.7.0.jar and hadoop-yarn-common-2.7.0.jar

Go to *grepcode.com* to see the source code (classes and interfaces) of any opensource software. For instance, search org.apache.hadoop.mapreduce. FileInputFormat in *grepcode.com*. You can see the source code and modify for your requirements. Suppose, if you want to override record reader, you can see a list of interfaces to implement, and classes to extend in *grepcode. com*. For a map and reduce task implementation, it is enough to import only mapreduce package. Similarly, if you want YARN services, then import yarn package in MR job.

HDFS API is available in org.apache.hadoop.fs package under a common category. This package contains classes such as FileSystem, FileStatus, FileUtil, Path, and interfaces such as PathFilter to work with input/output files. Command "hadoop fs" executes file system API internally. The HDFS libraries come under the package common because we can load files into HDFS even when MR and YARN services are not running. The HDFS is a storage part, and we can access HDFS files using a normal java program without any MR job. The MR uses this package internally to read and write files from HDFS. We can use MR with other file systems as well. Third party file systems are Amazon S3, Azure binary large object (blob), and Swift object storage. However, the native file system for Hadoop is HDFS. Blob, S3 are cloud storages, unlike HDFS. They are not as efficient as HDFS in a cluster environment.

6.8.1   WORDCOUNT PROGRAM IN MRV1 [12]

A job is any program written in a language. Today, MR jobs can be developed using different languages. A typical MR job comprises three functions: map, reudce, and driver (main function). The map function is included in Mapper class, reduce function is included in Reduce class, and driver function is included in the main class. An MR job can be developed in two ways.

1. Mapper class, Reducer class, and driver functions are included within another class.
2. Mapper class, Reducer class, and driver functions are created as an individual class.

Mapper class, Reducer class, and driver function are included within another class

```
import java.io.IOException;
import java.util.Iterator;
import java.util.StringTokenizer;
import org.apache.hadoop.conf.Configured;
import org.apache.hadoop.fs.Path;
import org.apache.hadoop.io.IntWritable;
import org.apache.hadoop.io.LongWritable;
import org.apache.hadoop.io.Text;
import org.apache.hadoop.mapred.FileInputFormat;
import org.apache.hadoop.mapred.FileOutputFormat;
import org.apache.hadoop.mapred.JobClient;
import org.apache.hadoop.mapred.JobConf;
import org.apache.hadoop.mapred.MapReduceBase;
import org.apache.hadoop.mapred.Mapper;
import org.apache.hadoop.mapred.Reducer;
import org.apache.hadoop.mapred.OutputCollector;
import org.apache.hadoop.mapred.Reporter;
import org.apache.hadoop.util.Tool;
```

```
import org.apache.hadoop.util.ToolRunner;
public class WC extends Configured implements Tool{
    public static class UDFmapper extends MapReduceBase
    implements Mapper<LongWritable, Text, Text, IntWritable> {
        private final static IntWritable one = new
        IntWritable(1);
        private Text word = new Text();
        public void map(LongWritable key, Text value,
        OutputCollector<Text, IntWritable> output, Reporter
        reporter) throws IOException {
            String line = value.toString();
            StringTokenizer tokenizer = new
            StringTokenizer(line);
            while (tokenizer.hasMoreTokens()) {
                word.set(tokenizer.nextToken());
                output.collect(word, one);
            }
        }
    }
    public static class UDFreducer extends MapReduceBase
    implements Reducer<Text, IntWritable, Text, IntWritable>
    {
        public void reduce(Text key, Iterator<IntWritable>
        values, OutputCollector<Text, IntWritable> output,
        Reporter reporter) throws IOException {
            int sum = 0;
            while (values.hasNext()) {
                    sum += values.next().get();
            }
            output.collect(key, new IntWritable(sum));
        }
    }
    public int run(String[] args) throws Exception {
        if(args.length <2) {
            System.err.println("Usage:<inputpath>
            <outputpath>");System.exit(-1);
        }
        JobConf job = new JobConf(WC.class);
        job.setJarByClass(WC.class);
        job.setMapperClass(UDFmapper.class);
        job.setReducerClass(UDFreducer.class);
```

```
        job.setOutputKeyClass(Text.class);
        job.setOutputValueClass(IntWritable.class);
        FileInputFormat.addInputPath(job,
        new Path(args[0]));
        FileOutputFormat.setOutputPath(job, new
        Path(args[1]));
        JobClient.runJob(job);
        return 0;
    }
    public static void main(String[] args) throws Exception {
        int exitCode = ToolRunner.run(new WC(), args);
            System.exit(exitCode);
    }
}
```

Mapper class, Reducer class, and driver function are created as individual class

Only the signatures are different between MRv1 and MRv2 programming construct. Please observe the words highlighted with bold face.

WCMRv1Driver.java

```
public class WCMRv1Driver{
    public static void main(String[] args) throws Exception {
        JobConf job = new JobConf(WCMRv1Driver.class);
        if(args.length <2) {
            System.err.println("Usage:<inputpath>
            <outputpath>"); System.exit(-1);
        }
        job.setJarByClass(WCMRv1Driver.class);
        job.setMapperClass(WCMRv1Mapper.class);
        job.setReducerClass(WCMRv1Reducer.class);
            job.setOutputKeyClass(Text.class);
            job.setOutputValueClass(IntWritable.class);
            FileInputFormat.setInputPaths(job, new
            Path(args[0]);
            FileOutputFormat.setOutputPath(job, new
            Path(args[1]);
            JobClient.runJob(job);
    }
}
```

WCMRv1Mapper.java

```
public class WCMRv1Mapper extends MapReduceBase implements
Mapper<LongWritable, Text, Text, IntWritable> {
    private final static IntWritable one = new IntWritable(1);
    private Text word = new Text();
    public void map(LongWritable key, Text value,
    OutputCollector<Text, IntWritable> output, Reporter
    reporter) throws IOException {
        String line = value.toString();
        StringTokenizer tokenizer = new StringTokenizer(line);
        while (tokenizer.hasMoreTokens()) {
            word.set(tokenizer.nextToken());
            output.collect(word, one);
            }
        }
    }
}
```

WCMRv1Reducer.java

```
public class WCMRv1Reducer extends MapReduceBase implements
Reducer<Text, IntWritable, Text, IntWritable>{
    public void reduce(Text key, Iterator<IntWritable> values,
    OutputCollector<Text, IntWritable> output, Reporter
    reporter) throws IOException {
        int sum = 0;
        while (values.hasNext()) {
            sum += values.next().get();
        }
        output.collect(key, new IntWritable(sum));
    }
}
```

6.8.2 WORDCOUNT PROGRAM IN MRV2 [13]

Mapper class, Reducer class, and driver function are included within another class

```
import java.io.IOException;
import java.util.StringTokenizer;
import org.apache.hadoop.conf.Configuration;
import org.apache.hadoop.fs.Path;
import org.apache.hadoop.io.IntWritable;
import org.apache.hadoop.io.LongWritable;
```

```
import org.apache.hadoop.io.Text;
import org.apache.hadoop.mapreduce.Job;
import org.apache.hadoop.mapreduce.Mapper;
import org.apache.hadoop.mapreduce.Reducer;
import org.apache.hadoop.mapreduce.lib.input.FileInputFormat;
import org.apache.hadoop.mapreduce.lib.output.
   FileOutputFormat;
public class WC{
    public static class UDFmapper extends Mapper<LongWritable,
    Text, Text, IntWritable>{
        private final static IntWritable one = new
        IntWritable(1);
        private Text word = new Text();
        public void map(LongWritable key, Text value, Context
        context)throws IOException, InterruptedException {
            String line = value.toString();
            StringTokenizer itr = new StringTokenizer(line);
    while (itr.hasMoreTokens()) {
        word.set(itr.nextToken());
        context.write(word, one);
        }
        }
    }
    public static class UDFreducer extends
    Reducer<Text,IntWritable, Text,IntWritable> {
        private IntWritable result = new IntWritable();
        public void reduce(Text key, Iterable<IntWritable>
        values, Context context) throws IOException,
        InterruptedException {
            int sum = 0;
            for (IntWritable val : values) {
                sum += val.get();
            }
            result.set(sum);
            context.write(key, result);
        }
    }
    public static void main(String[] args) throws Exception {
        Configuration conf = new Configuration();
        Job job = Job.getInstance(conf, "wordcount");
```

```
    job.setJarByClass(WC.class);
        job.setMapperClass(UDFmapper.class);
        job.setReducerClass(UDFreducer .class);\
        job.setOutputKeyClass(Text.class);
        job.setOutputValueClass(IntWritable.class);
        FileInputFormat.addInputPath(job, new
        Path(args[0]));
        FileOutputFormat.setOutputPath(job, new
        Path(args[1]));
        System.exit(job.waitForCompletion(true)? 0:1);
    }
}
```

Mapper class, Reducer class, and driver function are created as individual class

Only the signatures are different between MRv1 and MRv2 programming construct. Please observe the words highlighted with bold face.

WCMRv2Driver.java

```
public class WCMRv2Driver {
    public static void main(String[] args) throws Exception {
        if(args.length !=2) {
            System.err.println("Usage:<inputpath>
            <outputpath>");
            System.exit(-1);
        }
        Configuration conf = new Configuration();
        Job job = Job.getInstance(conf, "wordcount");
        job.setJarByClass(WCMRv2Driver.class);
        job.setMapperClass(WCMRv2Mapper.class);
        job.setReducerClass(WCMRv2Reducer.class);
        job.setOutputKeyClass(Text.class);
        job.setOutputValueClass(IntWritable.class);
        FileInputFormat.addInputPath(job, new Path(args[0]);
        FileOutputFormat.setOutputPath(job,new Path(args[1]));
        System.exit(job.waitForCompletion(true) ? 0 : 1);
    }
}
```

WCMRv2Mapper.java

```
public class WCMRv2Mapper extends Mapper<Object, Text, Text,
    IntWritable>{
    private final static IntWritable one = new IntWritable(1);
    private Text word = new Text();
    public void map(Object key, Text value, Context context)
    throws IOException, InterruptedException {
        String line = value.toString();
        StringTokenizer tokenizer = new StringTokenizer(line);
        while (tokenizer.hasMoreTokens()) {
            word.set(tokenizer.nextToken());
            context.write(word, one);
        }
    }
}
```

WCMRv2Reducer.java
```
public class WCMRv2Reducer extends
Reducer<Text,IntWritable,Text, IntWritable>{
    private IntWritable result = new IntWritable();
    public void reduce(Text key, Iterable<IntWritable> values,
    Context context) throws IOException, InterruptedException
    {
        int sum = 0;
        for (IntWritable val : values) {
            sum += val.get();
        }
        result.set(sum);
        context.write(key, result);
    }
}
```

6.8.3   *DIFFERENCES BETWEEN OLD AND NEW API*

The purpose of the new API is better readability and to fix a few issues to adapt YARN. However, the major intention is to reduce the number of interfaces defined in MRv1, which forces us to override all methods. In MRv2, they have been redefined as abstract classes. Therefore, only the required methods are overridden, and new concrete methods can be added. YARN supports programs written in both older API and a new API. Although Hadoop is shipped with both old and new MR APIs, signatures of methods are not the same. Before proceeding further, did you observe the signatures bolded in the last couple of pages for MRv1 and MRv2 jobs? If yes, let us continue reading further. The differences are:

old API	new API
In driver function	
1. org.apache.hadoop.mapred package	org.apache.hadoop.mapreduce package
2. JobConf object is used to configure job	Job object is used to configure job
3. JobClient.runJob() to submit job to JT	job.waitForCompletion(true) submits job
In Mapper class	
4. Extends MapReduceBase and implements Mapper interface	Extends Mapper abstract class
5. Throws only IOException	Throws IOException, and InterruptedException
6. OutputCollector, Reporter	Both are done by Context class
7. collect(k, v) writes into memory buffer	write(k,v) writes into memory buffer
In Reducer class	
8. Extends MapReduceBase and implements Reducer interface	Extends Reducer abstract class
9. Throws only IOException	Throws IOException, and Interrupted Exception
10. OutputCollector, Reporter	Both are done by Context class
11. Iterator is used for enumeration	Iterable is used for enumeration
12. while(values.hasNext())	for-each loop to iterate values
13. collect(k, v) writes into memory buffer	write(k,v) writes into memory buffer
Other differences	
14. Partitioner is an interface	Partitioner is an abstract class
15. IdentityMapper, IdentityReducer	Default classes are Mapper and Reducer
16. Default block size is 64 MB	Default block size is 128 MB
17. We can set the number of map tasks using set NumMapTasks(int). It is set based on the	No such methods. Number of map tasks are calculated based on the number of IS. number of IS internally.
18. _logs file is created in output directory	_logs is not created in the output directory.

Context unifies the role of JobConf, OutputCollector, and Reporter from old API. User-overridable methods in the new API are declared to throw InterruptedException. This means that you can write your code to be responsive to interrupts so that the framework can gracefully cancel long-running operations if it needs to. Jobs using new API that was compiled against Hadoop v1 need to be recompiled to run against Hadoop v2. This is because some classes in the new MR API have been changed to interfaces. If it is not compiled with the right version, it gives runtime error as below.

```
java.lang.IncompatibleClassChangeError:
Found interface org.apache.hadoop.mapreduce.
TaskAttemptContext, but class was expected
```

When converting old MR program to new API, do not forget to change the signatures of the map() and reduce() methods to the new form. Just changing your class to extend the new Mapper or Reducer classes will not produce a compilation error or warning, because these classes provide a default map() and reduce() methods which will override your map() and reduce() methods. It is very hard to diagnose such problems. However, annotating your map() and reduce() methods with the @override annotation will allow the java compiler to catch these errors.

6.8.4   HADOOP STREAMING API AND HADOOP PIPES

The MR jobs can be developed in many languages: Java, Python, Ruby, C++, etc. However, the only requirement of a language is, it should be able to read and write from standard input and output. There is a framework called Jython that converts python programs into a jar and allows you to run as a normal java job. However, as shown in Figure 6.5, Hadoop provides streaming API that lets you run Python, Ruby, etc. directly on Hadoop without converting into the jar. C++ uses Hadoop pipes to launch MR job. Therefore, the same data can be processed by different languages in a distributed environment. Let us write an MR wordcount job in python.

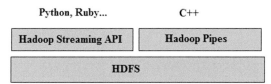

FIGURE 6.5   Hadoop Streaming and Pipes.

Create mapper and reducer in python [14]

```
$ vi mapper.py
    #!/usr/bin/python
    import sys
    for myline in sys.stdin:          # Mapper input is taken from standard input
        myline= myline.strip()        # Remove whitespace either side
        words = myline.split()        # Break the line into words
        for myword in words:          # Iterate the words list

    print '%s\t%s' % (myword, 1) # Write mapper results to standard output
$ vi reducer.py
    #!/usr/bin/python
    from operator import itemgetter
    import sys
    current_word = None
    current_count = 0
    word = None
    for myline in sys.stdin:          # Reducer input is taken from standard
    input
        myline = myline.strip()       # Remove whitespace either side
        word, count = myline.split    # Split the input we got from mapper.py
        ('\t', 1) try:                # Convert count variable to integer
            count = int(count)
    except ValueError:
            continue                  # Count was not a number, so silently
                                      ignore this line
    if current_word == word:
        current_count += count
    else:
    if current_word:                  # Write result to standard output
            print '%s\t%s' % (current_word, current_count)
    current_count = count
    current_word = word
    if current_word == word:          # Do not forget to output the last word if
                                          needed!
            print '%s\t%s' % (current_word, current_count)
```

Run python mapper and reducer tasks on the local file system

$ echo "hi how are you" | python mapper.py

```
hi          1
how         1
are         1
you         1
```

$ echo "hi how are you hi" | python mapper.py | sort -k1,1 | python reducer.py

```
are         1
hi          2
how         1
you         1
```

Run python mapper and reducer tasks on HDFS

• Make sure HDFS and YARN runs (either single node or multi-node)
• Download hadoop-streaming-2.7.3.jar.
• Upload input file onto HDFS

  $ hdfs dfs -put input.txt /

• Run python MR job: syntax is

  $ yarn jar hadoop-streaming-2.7.3.jar -file mapper_location -mapper mapper.py -file reducer_location -reducer reducer.py -input /hdfs_input -output /hdfs_output

  $ yarn jar hadoop-streaming-2.7.3.jar -file mapper.py -mapper mapper.py -file reducer.py -reducer reducer.py -input /input.txt -output /python_output

6.9   ART OF WRITING MR JOBS

Driver function consists of a block of codes that control the execution sequence of MR jobs. We need to understand the purpose of each line and where it should be placed in the code block. It comes by practice to precisely use. However, we have given some ideas on driver function, Mapper class, and Reducer class. Let us see what we need to consider while developing an MR job and preparing dataset.

6.9.1 BEFORE DEVELOPING MR APPLICATION

Before dirtying your hands with java program for MR job, you have to do a hand-sketch on certain key-parameters that will help you in picking up right data type for key-value pair and user-defined function apart from the map and reduce function. Follow the sequence of steps given below to sketch what you may require while developing MR job. Please refer Section 2.7 and Section 2.8 to understand better while reading the steps given below.

1. Set up a single node or multi-node installation using a cluster of PMs/VMs.
2. Understand the type of dataset you have. Example: Text or image, etc.
3. Determine the block size you want to load dataset onto HDFS.
4. Define the problem statement.
5. Determine the file input format for Mapper class based on your dataset. Example: Text InputFormat, SequenceFileInputFormat, etc.
6. If you define your file input format, then customize its IS and RR. However, this is optional.
7. Define a new Mapper class if there is no suitable predefined Mapper for your dataset.
8. Identify map input key, map input value, map output key, and map output value for Mapper class and determine their datatype.
9. In the map, perform input parsing, projection (selecting relevant fields), and filtering records that are not of interest. Even though the MR framework itself decides the number of map tasks, you can decide it by changing the block size while loading dataset or changing IS size while launching a job.
10. Identify whether we have the advantage of using combiner.
11. Define Reducer class if there is no suitable predefined Reducer class for the problem statement.
12. Identify reduce input key, reduce input value, reduce output key, and reduce output value for Reducer class and determine their datatype.
13. Implement core algorithm concepts in reduce function.
14. Determine output file format to write output key-value pair.

15. You can determine how many numbers of reduce tasks to use to exploit parallelism based on your observation on map output size.

16. Identify whether to use predefined/user-defined partitioner to balance the load across reduce tasks.

6.9.2   DRIVER CLASS STEPS

While writing driver function (main function), we have to follow the order given below to be away from error. Refer MR wordcount programs given in Sections 6.8.1 and Section 6.8.2 while reading the steps given below for better understanding.

1. Check for minimum command line arguments.
2. Create an object for Configuration class.
3. Create an object for the Job class.
4. Set the jar by class name.
5. Set up Mapper and Reducer class. Combiner and partitioner are optional unless you plan for it.
6. Set up data type for output key and output value for Mapper and Reducer class.
7. Add input and output file format.
8. Add the path for the input file and output directory.
9. Submit job using waitForCompletion().

6.9.3   DEBUGGING MR JOBS

In order to debug a job in a distributed environment, you can use a print statement in map/reduce tasks. However, the print statement displays information only in the command line of the respective slave node where a particular task is running. Therefore, it is wise to log the activities in files and review later. We can also use setStatus() method. Counters are another manual way of collecting job status for diagnosing the MR jobs.

6.9.3.1   LOGS

Logs are a good friend for troubleshooting. It records events such as error messages, exceptions. All Hadoop daemons produce log files to record what is happening in the system. You can see these logs in WUI of NN and JT. If any problem in running Hadoop services, you can track these logs to spot the problem. In general, there are two log files: system logs and application (Hadoop) logs.

System log files

System log files are maintained in /var/log by default. Log location can be changed using HADOOP_LOG_DIR in hadoop-env.sh.

```
$ vi /usr/local/hadoop/etc/hadoop/hadoop-env.sh
    export HADOOP_LOG_DIR=/var/log/hadoop
```

The log directory will be created if it does not already exist.

Hadoop log files

Each Hadoop daemon running in a node produces two log files: log and out. These files are in /usr/local/hadoop/logs location.

Log – is written by the log4j tool. This file, whose name ends in .log, should be the first try when diagnosing problems because application-related log messages are written here.

Out – is the combined standard output and standard error log. This log file name ends with .out, usually contains little or no output since Hadoop uses log4j for logging.

Only the last five log files are retained. Log files are suffixed with a number between 1 and 5, with 5 being the oldest file. Log file name comprises various information: hadoop-username-deamon-hostname.log/out.number. Example: hadoop-itadmin-datanode-ubuntu.log.1

Go to /usr/local/hadoop/logs or browse http://NN_IP:50070/logs/

```
hadoop-itadmin-secondarynamenode-ubuntu.out.4
hadoop-itadmin-secondarynamenode-ubuntu.out.5
mapred-itadmin-historyserver-ubuntu.log
mapred-itadmin-historyserver-ubuntu.out
mapred-itadmin-jobhistoryserver-ubuntu.out
SecurityAuth-itadmin.audit
userlogs
yarn-itadmin-jobhistoryserver-ubuntu.out
yarn-itadmin-jobhistoryserver-ubuntu.out.1
yarn-itadmin-nodemanager-ubuntu.log
yarn-itadmin-nodemanager-ubuntu.out
```

```
yarn-itadmin-nodemanager-ubuntu.out.1
yarn-itadmin-nodemanager-ubuntu.out.2
```

Log4j property controls the log file location and its name. You can customize log4j properties

$ vi /usr/local/hadoop/etc/hadoop/log4j.properties

Container writes stdout and stderr in local file system specified by yarn. nodemanager.log-dirs/ <applications_id>/<container_id> in NM itself. Once application completed, if aggregation is enabled (yarn.log-aggregation-enable as true), all container logs are copied into HDFS by JHS. Local logs in NM are deleted after yarn.nodemanager.delete.debug-delay-sec. To view application log:

$ yarn application -list
$ yarn logs -applicationId application_1496656498736_0004

To see local logs of NM, go to NM_IP:8042 in the browser → go to Tools → local logs. To access MRAppMaster logs, go to RM_IP:8088 in the browser → cluster. It displays the URI of MRAppMaster (also displayed on the console while launching job). Once the job is done, logs-related to MRApp-Master and tasks are moved to JHS. You can use mapred job -logs command to view them. The reason why RM does not maintain container ID is, to make RM lightweight with less meta-data in memory. Link to the MRAppMaster alone is maintained in RM, not the meta-data of each container. Hence, RM does not keep track of containers. MRAppMaster itself takes containers care.

Compressing aggregated logs is disabled by default. To enable, set lz4 or gzip to yarn.node manager.log-aggregation.compression-type. Logs in the local file system in NM is retained for 10800 seconds (3 hours) using yarn. nodemanager.log.retain-seconds property. If log aggregation is enabled, retain option is disabled, because once the logs are copied into HDFS, they are deleted in NM. However, deletion also can be set using the property yarn.node manager.delete.debug-delay-sec in seconds.

Whatever stored in HDFS is stored as blocks with replication. Therefore, meta-data for log file such as name, block ID, etc. have to be maintained in NN memory. Aggregating log data for 1000s of jobs leads to NN memory exhausted. Therefore, to manage log files more extensively, we can use external tools such as Loggly, Splunk, Sumo Logic, Sematext, LogStash, GrayLog, PaperTrails, Hunk. Hunk provides support for aggregating logs from both Hadoop v1 and v2. It comes with a lot of predefined queries, visualization, monitoring features, etc.

Audit logging – The HDFS can log all file system access requests using log4j at INFO-level. By default, it is disabled. To enable, include the following line in hadoop-env.sh:

$ sudo vi /usr/local/hadoop/etc/hadoop/hadoop-env.sh

```
export HDFS_AUDIT_LOGGER="INFO,RFAAUDIT"
```

Getting stack traces – Hadoop daemons produce a thread dump for all running threads in the JVM itself. This can be accessed via WUI. For instance, a thread-dump for RM can be seen in http://RM_IP:8088/stacks.

6.9.3.2 COUNTERS

The counter is a useful feature provided by Hadoop and helps to gather statistics of jobs across MR phases to determine data quality, diagnose, and debug problems. For instance, you can calculate the number of bad records in your dataset that is poorly formed. Therefore, it is good to use counters rather than logging at times. Because it is easy to retrieve and display counter values than viewing logs and read it to spot the problems for large distributed jobs. Counters can be accessed via WUI and CLI, sometimes using APIs in normal java program. There are two types of counters: built-in and user-defined counters.

Built-in counters

The MR framework provides a set of built-in counters for every job to report various information. These counters are classified into several groups: MR task counters, Job counters, File System Counters, FileInputFormat counters, FileOutputFormat counters.

1. MR task counters: Task counters are updated as the task progresses. YARN child of each task increments task counters such as the number of records processed by map/reduce task and sends summary to MRAppMaster, which consolidates and provides the details of input/output records/bytes processed by map/reduce tasks, materialized bytes, shuffle bytes, spilt records, CPU time in milliseconds, physical/virtual memory consumed in bytes, etc.

2. Job counters: Job counters are updated as job progress and maintained by MRAppMaster. It monitors information such as the number of map/reduce tasks launched, data local tasks, number of uber tasks, killed/failed tasks, etc.

3. File system counters: Number of bytes read/written, number of read/write operations in disks for each file system (not only HDFS and also for the local file system, S3).

4. FileInputFormat counters: Number of bytes read by map tasks using FileInputFormat.

5. FileOutputFormat counters: Number of bytes written by reduce tasks using FileOutput Format.

User-defined counters

Users can create their counters to enumerate something of interest in map/reduce tasks. It can be effectively utilized to monitor tasks across nodes and created in two ways: enum and dynamic counters (without enum).

*   In enum, enum name is a group name and enum fields are counter names. The MR framework turns enums into strings to send counters over RPC and to manage more easily.
*   You cannot create counters on the fly using enums, because java enums are defined at compile time. Dynamic counters are created with the help of the Context object. These user-defined counters are global, and the MR framework aggregates them across all map/reduce tasks to produce total summary at the end of the job.

Once a job is completed, MRAppMaster displays group name and counter name with collected data on the console as follows.

```
File System Counters
    FILE: Number of bytes read=4929778
    FILE: Number of bytes written=10667268
    FILE: Number of read operations=0
    FILE: Number of large read operations=0
    FILE: Number of write operations=0
    HDFS: Number of bytes read=161235975
    HDFS: Number of bytes written=1500726
    HDFS: Number of read operations=21
Job Counters
    Killed map tasks=1
    Launched map tasks=7
    Launched reduce tasks=1
    Data-local map tasks=7
    Total time spent by all maps in occupied slots
    (ms)=182357
    Total time spent by all reduces in occupied slots
    (ms)=8764
```

```
    Total time spent by all map tasks (ms)=182357
    Total time spent by all reduce tasks (ms)=8764
    Total vcore-seconds taken by all map tasks=182357
    Total megabyte-seconds taken by all map tasks=186733568
    Total megabyte-seconds taken by all reduce
    tasks=8974336
  Map-Reduce Framework
    Map input records=3471692
    Map output records=3471692
    Map output bytes=97815553
    Map output materialized bytes=4929808
    Combine input records=3471692
    Combine output records=163345
    Reduce input groups=163345
    Reduce shuffle bytes=4929808
    Reduce input records=163345
    Reduce output records=163345
    Spilled Records=326690
    Shuffled Maps =6
    Merged Map outputs=6
    GC time elapsed (ms)=5620
    CPU time spent (ms)=184490
```

Job history information is persistently stored in HDFS by JHS. Location can be set using mapreduce.jobhistory.done-dir property. All this history of information is stored in JSON format. It can be accessed via WUI of JHS. You can access these counters using simple java program with appropriate API. It can also be accessed on the command line using the following command:

$ mapred job -counter < job-id> <group-name> <counter-name>

Metrics vs Counters

Hadoop daemons collect information (also known as metrics) about events. Metrics belong to contexts such as dfs, mapred, yarn, and rpc. Each Hadoop daemon usually collects metrics under several contexts. Example: DNs collect metrics such as the number of bytes written, number of blocks replicated, and the number of read requests from clients. How do metrics differ from counters?

- The main difference is their scope: metrics are collected by Hadoop daemons, whereas counters are collected for MR tasks and aggregated for the whole job.
- They have different audiences: metrics for administrators, and counters for MR users.

- The way they are collected and aggregated is also different. Counters are MR feature and propagated (via RPC) to MRAppMaster from the task JVM. The collection mechanism for metrics is decoupled from the MR component that receives the updates. It uses various pluggable outputs, including Ganglia, and JMX. The daemon collecting the metrics performs aggregation before they are sent to the output.

6.9.3.3   *VERSION MISMATCH EXCEPTION*

After executing "namenode -format" command in NN, NamespaceID, clusterID, blockpoolID are created in /usr/local/hadoop/tmp/dfs/name/current/VERSION, which denotes the current session of Hadoop cluster. The NN will regenerate all these IDs upon reformatting. DNs maintain NamespaceID in /usr/local/hadoop/tmp/dfs/data/current/VERSION to synch with NN. If the NN is reformatted, DNs cannot sync with NN as DNs have old NamespaceID. Therefore, we have to keep NamespaceID synch in NN and DN, else version mismatch will be the error that leads to loss of data blocks. Similarly, blockpoolID is maintained in all DNs in the following location: /usr/local/hadoop/tmp/dfs/data/current/. BlockpoolID in NN and all DNs must be the same. To see the list of blocks stored in DNs,

```
$ ls /usr/local/hadoop/tmp/dfs/data/current/BP-1390157978-
10.100.52.146-1491543068972/current/finalized/subdir0/
subdir0/
```

```
blk_1073741825      blk_1073741827      blk_1073741829
blk_1073741825_1001.meta    blk_1073741827_1003.meta
blk_1073741829_1005.meta
blk_1073741826      blk_1073741828      blk_1073741830
```

You can see these IDs in NN WUI. If you do not handle version mismatch exception, you will have to lose all the data uploaded in HDFS. However, you can track such exception in the following location: /usr/local/hadoop/logs/hadoop-username-datanode-slave1.log which contains an entry error incompatible namespace. After replacing the correct IDs in all DNs, restart the Hadoop services using start-dfs.sh. Now, hadoop fsck command will display block information. Alternatively, you can delete tmp directory in NN and all DNs, and reformat HDFS. However, you will lose all existing data.

Note: If there are problems in starting Hadoop services, always first check the XML configuration files. Mostly the problem will be with hostname mismatch, IP address misconfiguration, etc.

6.9.4   TUNING MR JOBS TO IMPROVE PERFORMANCE

In this section, we shall discuss some important properties that help improve performance, skipping bad records, condition for launching reduce tasks, and the problem with a large number of small files.

6.9.4.1   HADOOP CONFIGURATION PROPERTIES

Despite writing efficient and effective MR programs, unless configuring MR properties properly, there is no chance of improving MR performance and resource utilization. For instance, you can observe the decrease in the number of mappers if you change the default block size 64 MB to 128 MB. Therefore, more jobs can run simultaneously. However, the latency and IO activity of mapper will be high. Following are the list of parameters, for which you may identify some suitable value to improve the MR performance.

- Block size, IS size, RF.
- Number of map and reduce tasks.
- Combiner, partitioner, compression scheme, job scheduler.
- Intermediate buffer size, spill size, the number of threads to copy map output.
- HDD and network data transfer rate.
- Container and JVM configuration for the map and reduce tasks.

Tuning map-side properties

mapreduce.task.io.sort.mb – Memory buffer to write map output. The default value is 100 MB.

mapreduce.map.sort.spill.percent – Amount of data to be filled in in-memory before spilling into a disk. The default value is 0.80.

mapreduce.task.io.sort.factor – Maximum number of spills to be merged while sorting files. This property is used in the reducer. The default value is 10. It is fairly common to increase up to 100.

mapreduce.map.combine.minspills – Minimum number of spill files needed for the combiner to run (if combiner is specified). The default value is 3.

mapreduce.map.output.compress – to compress map output. The default value is false.

mapreduce.map.output.compress.codec–Compression codec to use for the map outputs. The default value is org.apache.hadoop.io.compress.DefaultCodec.

mapreduce.shuffle.max.threads – Number of worker threads per NM for serving map output to reducers. This cannot be set for individual jobs. The default value is 0.

Tuning reduce-side properties

mapreduce.reduce.shuffle.parallelcopies – Number of threads used by reducer to copy map output. The default value is 5.

mapreduce.reduce.shuffle.maxfetchfailures – Number of times a reducer tries to fetch map output before reporting an error. The default value is 10.

mapreduce.reduce.shuffle.input.buffer.percent – Proportion of the total heap size to be allocated for map output buffer during the copy phase in the shuffle. The default value is 0.7.

mapreduce.reduce.shuffle.merge.percent – Threshold usage proportion for the map output buffer (mapred.job.shuffle.input.buffer.percent) for starting the merging process and spilling to disk. The default value is 0.66.

mapreduce.reduce.input.buffer.percent – Proportion of the total heap size to be used for retaining map output in the in-memory buffer during the shuffle. The default value is 0.0.

6.9.4.2 *SKIPPING BAD RECORDS AND FAILURE PERCENTAGES*

Failure of the map/reduce task can happen due to a variety of reasons: bugs in map/reduce coding, bad input records, and bugs in third-party libraries. In the first case, it is better to debug the code, find the cause of failures, and fix it. We code an MR job to process data records in a certain fashion. Suppose input dataset is not well-formed, the program can cause an exception and exit the task. A record that causes failure is called a bad record. Bad records can happen due to an ill-formed dataset or block corruption. Therefore, either you have to handle exceptions raised due to bad records or frame the dataset in such a way that it does not raise exceptions. In the third case, it is better to choose the right third-party jars to support the MR jobs.

In the second case, if there is a record that causes an exception, the map task is re-attempted. Tasks can be re-attempted up to four times by default (mapreduce.map/reduce.maxattempts). If a task is terminated continuously three times due to bad record, the MR framework skips the bad record and takes the next record in the fourth attempt of map task for the same block in the same node. Skipping bad records during the map and reduce task is enabled by default. To skip bad records,

- mapreduce.map.skip.maxrecords (default 0) for map task.
- mapreduce.reduce.skip.maxgroups (default 0) for reduce task.

However, it is not the best solution if there are 1000s of bad records, causing a large number of map/reduce task re-attempts. Map/reduce task will be terminated if it neither reads nor writes for 10 minutes by default (mapreduce.task.timeout). When processing a large amount of data, there may be cases where a small number of map/reduce tasks will fail, but the final result makes sense. Therefore, it is possible to continue job running even if some small percentage of map/reduce tasks have failed.

- mapreduce.map.failures.maxpercent (default 0) for map task.
- mapreduce.reduce.failures.maxpercent (default 0) for reduce task.

If this property is assigned with 50, map/reduce task will complete processing even if 50% of map/reduce tasks have failed. You have to decide the % of failed tasks to skip according to the application requirement.

6.9.4.3   WHEN SHOULD REDUCE TASKS BE LAUNCHED?

In general, reduce phase progress is split into three parts:

0–33%	– shuffle completed
34–66%	– merge+sort+group completed
67–100%	– the number of reduce tasks completed. If only half the number of reduce tasks are completed, then it is $1/3+1/3+1/6=83\%$.

Sort in reduce phase can start only after all reduce tasks received inputs from map tasks and passed shuffle+merge. At times, reducer progress seems stuck at 33%. Because shuffle is waiting for all map tasks to finish. You can customize to start reduce tasks early or later using the property mapreduce. job.reduce.slowstart.completedmaps.

- A value of 1.0 will wait for all the map tasks to finish before sorting in reduce phase begins.
- A value of 0.0 will start all reduce tasks right after job launched along with all map tasks.
- A value of 0.5 will start the reduce tasks when half the number of map tasks are completed.

You can also change this per job basis. Why is starting reduce tasks early a good thing? Because all map tasks start sending the intermediate output to reduce tasks without spilling into local disk. It is a good thing if your network is the bottleneck and do not want to increase the traffic. However, it clogs up the cluster resource for a long time. It reduces the opportunity of running more jobs at the same time.

6.9.4.4   *LARGE NUMBER OF SMALL FILES LESS THAN THE BLOCK SIZE*

If there are too many small files (like image files) less than the block size, arbitrary file size itself is considered to be block size. This results in huge meta-data (each object such as a file, directory, and block in HDFS takes 150 bytes for meta-data) in memory. If there are 10 million small files (images), then its meta-data size is about 3 GB in NN memory. Each small block is processed by a separate map task. Therefore, a huge number of map tasks have to be launched. This leads to a smaller number of jobs running at the same time due to resource unavailability. There are ways to combine many small files up to a default block size, so a smaller number of map tasks are launched. For instance, let us create six small files and discuss with possible solutions to handle this situation.

```
$ mkdir data
$ cd data
$ vi file1
      hi how are you
$ vi file2
      how is your brother
$ vi file3
      what was your lunch
```

```
$ vi file4
        how is your studies
$ vi file5
        how is your bike
$ vi file6
        what is your aim
$ cd ..
$ hdfs dfs -put data /
$ hdfs dfs -lsr /data
    -rw-r--r-- 3 itadmin supergroup 16 2017-04-19 17:56
    /data/file1
    -rw-r--r-- 3 itadmin supergroup 19 2017-04-19 17:56
    /data/file2
    -rw-r--r-- 3 itadmin supergroup 23 2017-04-19 17:56
    /data/file3
    -rw-r--r-- 3 itadmin supergroup 18 2017-04-19 17:56
    /data/file4
    -rw-r--r-- 3 itadmin supergroup 17 2017-04-19 17:56
    /data/file5
    -rw-r--r-- 3 itadmin supergroup 17 2017-04-19 17:56
    /data/file6
```

Solution 1: set IS size bigger

Many small blocks are put into an IS. Changing block size after data stored in HDFS is a time-consuming process. Therefore, set block size in advance wisely based on your application when you upload data onto HDFS. However, you can change the IS size at run time. You can set the IS size (in bytes) in driver function as given below. Default value 0 means that block size itself is IS size.

```
job.set("mapreduce.input.fileinputformat.split.maxsize", "64000000");
job.set("mapreduce.input.fileinputformat.split.minsize", "64000000");
```

alternatively, you can override via command line using -D at run time.

```
$ yarn jar job.jar classname -D mapreduce.input.fileinputformat.split.
minsize=64000000 /data /output
```

alternatively, via conf file (you must use GenericOptionsParser to read conf from command line arguments)

```
$ vi conf.xml
```

```
<configurations>
   <property>
      <name> mapreduce.input.fileinputformat.split.
      maxsize </name>
      <value> 64000000</value>
   </property>
   <property>
      <name> mapreduce.input.fileinputformat.split.
      minsize </name>
      <value> 64000000</value>
      <property>
</configurations>
```

$ yarn jar job.jar classname -conf conf.xml /data /output

Once the job is done, traverse the job metrics/counters in the console to find out the number of IS formed. See the job counter to find the number of map tasks, the number of data local map tasks, rack local map tasks, etc. If data locality is < 98%, there is some problem in the Hadoop cluster, because rack local map tasks should be >98%. You must consider the heap size in a container while setting a maximum IS size. If you set IS size beyond the block size, there will be an increase in the number of non-local execution, which ultimately impacts network performance. Example: IS comprises three blocks which are in different nodes. Now, the mapper is launched in any one of the nodes (say node1) to achieve data locality, and other two blocks are copied to node1 for processing. It takes more network bandwidth, deserialization/ serialization, compression/ decompression, etc. Therefore, to achieve over 98% data local map execution, it is always wise to set IS size as the block size (by default). The IS size can be determined by using the following formula:

```
max(minimum split size, min(maximum split size,
block size))
```

Solution 2: use CombineFileInputFormat

It creates only one IS by dumping all small files. It may involve more network traffic due to non-local execution. To achieve this, you have to set in the driver function as follows:

```
job.setInputFormatClass(CombineTextInputFormat.
class);
```

Solution 3: merge all small files or use the append command

You can merge all small files into one large file before you upload onto HDFS. So, the number of blocks will be less.

```
$ cat data/* >> mergedfile
$ hdfs dfs -put mergedfile /
```

alternatively, use the append command while uploading all small files onto HDFS.

```
$ hdfs dfs -appendToFile data/* /mergedfile
$ hdfs dfs -cat /mergedfile
```

```
hi how are you
how is your brother
what was your lunch
how is your studies
how is your bike
what is your aim
```

```
$ yarn jar /usr/local/hadoop/share/hadoop/mapreduce/hadoop-mapre-
duce-examples-2.7.0.jar wordcount /mergedfile /output
```

However, trying this approach for 1000s of files is a time-consuming process.

Solution 4: Hadoop archive

Using the Hadoop archive, you can dump many small files into a larger one, so that a smaller number of map tasks will run.

```
$ cat data/* >> mergedfile
$ yarn jar /usr/local/hadoop/share/hadoop/mapreduce/hadoop-mapre-
duce-examples-2.7.0.jar wordcount /data /output
```

Once this job is done, you can check the job counter in the console. The number of IS and map tasks launched will be 6 as we have 6 small files in the data directory. Now, let us work out with Hadoop archive to minimize the number of IS, thus minimizing the number of map tasks.

```
$ hadoop arhive
$ hadoop archive -archiveName file.har -p /hdfs_input_dir /
hdfs_output_location
$ hadoop archive -archiveName mergedfile.har -p /data /new
```

Hadoop archive runs as an MR job to combine all small files that are in the HDFS data directory and produces one large file in the target HDFS location.

$ hdfs dfs -ls /new

```
drwxr-xr-x - itadmin supergroup 0 2017-04-19 17:59 /
new/mergedfile.har
```

$ hdfs dfs -lsr /new/mergedfile.har (or)
$ hdfs dfs -lsr har:///new/mergedfile.har

```
-rw-r--r-- 3 itadmin supergroup 0 2017-04-19 17:59 /
new/mergedfile.har/_SUCCESS
-rw-r--r-- 5 itadmin supergroup 475 2017-04-19 17:59
/new/mergedfile.har/_index
-rw-r--r-- 5 itadmin supergroup 23 2017-04-19 17:59
/new/mergedfile.har/_masterindex
-rw-r--r-- 3 itadmin supergroup 110 2017-04-19 17:59
/new/mergedfile.har/part-0
```

`_SUCCESS`   denotes that the job is done successfully. This file size is 0.
`_masterindex`   helps to look up for data in the concatenated file.
`part-0`   indicates the concatenated file.

$ hdfs dfs -cat /new/mergedfile.har/part-0

```
hi how are you
how is your brother
what was your lunch
how is your studies
how is your bike
what is your aim
```

If you run a wordcount job on the archived file, it will launch one map task for each file.

$ yarn jar /usr/local/hadoop/share/hadoop/mapreduce/hadoop-mapreduce-examples-2.7.0.jar wordcount /new/mergedfile.har /output //number of IS is 6

Therefore, run wordcount job on part-0 file.

$ yarn jar /usr/local/hadoop/share/hadoop/mapreduce/hadoop-mapreduce-examples-2.7.0.jar wordcount /new/mergedfile.har/part-0 /output //number of IS is 1

HAR limitations:

- There is no support for compression. Therefore, the size of the archived file is equal to the size of the original files.

- Hadoop archive is immutable. Therefore, if you want to change in it, you have to re-create the archive file.

Solution 5: SequenceFileInputFormat

SequenceFile is a flat file consisting of binary key/value pairs and extensively used in MR as file input/output format. For instance, the temporary output of map tasks is stored as a sequence file internally. In HDFS, the sequence file is one of the solutions to small file problems. The concept of the sequence file is to merge small files into one large file. Example: suppose there are 100 KB sized 10,000 files, we can write a program to put them into a single sequence file. Here, the new sequence file name is the key, and the file content is value. Sequence file is splittable and supports compression. Block compression is generally the preferred option when using SequenceFile. It is good to use as file output format when the chain of MR jobs is used.

6.10   HADOOP MR BENCHMARK FOR CLUSTER PERFORMANCE TEST

Benchmark jobs are a way to verify HDFS and YARN performance in the cluster. You can download over 500 GB of data from Ref. [21] and test with some sample jobs given to measure the performance of the cluster. There are different performance-measuring benchmark jobs.

- teragen creates sample data.
- terasort sorts huge dataset.
- teravalidate ensures that terasort are mapped and reduced correctly.
- testDFSIO tests the input output performance of HDFS.
- nnbench checks the NN hardware.
- mrbench runs many small jobs.

Generate huge random text file with the size you wanted. Specify the size in bytes and I generate 5 GB.

$ yarn jar /usr/local/hadoop/share/hadoop/mapreduce/hadoop-mapreduce-examples-2.7.0.jar teragen 5000000000 /terasort-input **or**

$ yarn jar /usr/local/hadoop/share/hadoop/mapreduce/hadoop-mapreduce-examples-2.7.0.jar randomtextwriter -D mapreduce.randomtextwriter.totalbytes=5000000000 /terasort-input

Sort huge data

$ yarn jar /usr/local/hadoop/share/hadoop/mapreduce/hadoop-mapreduce-examples-2.7.0.jar terasort /terasort-input /terasort-output

Validating sort job

$ yarn jar /usr/local/hadoop/share/hadoop/mapreduce/hadoop-mapreduce-examples-2.7.0.jar teravalidate /terasort-output /terasort-validate

Run TestDFSIO to test for IO throughput of the cluster.

    -write: creates sample files
    -read: reads sample files
    -clean: deletes the test outputs
    -fileSize: specified with MBs

To test write performance

$ yarn jar /usr/local/hadoop/share/hadoop/mapreduce/hadoop-mapreduce-client-jobclient-2.7.0.jar TestDFSIO -write -nrFiles 10 -fileSize 100

```
17/04/30 19:32:59 WARN hdfs.DFSClient:
DFSInputStream closed already
17/04/30 19:32:59 INFO fs.TestDFSIO: ----- TestDFSIO
----- : write
17/04/30 19:32:59 INFO fs.TestDFSIO: Date time: Sun
Apr 30 19:32:59 IST 2017
17/04/30 19:32:59 INFO fs.TestDFSIO: Number of
files: 10
17/04/30 19:32:59 INFO fs.TestDFSIO: Total MBytes
processed: 1000.0
17/04/30 19:32:59 INFO fs.TestDFSIO: Throughput mb/
sec: 124.20817289777668
17/04/30 19:32:59 INFO fs.TestDFSIO: Average IO rate
mb/sec: 133.392333984375
17/04/30 19:32:59 INFO fs.TestDFSIO: IO rate std
deviation: 41.25280387986924
17/04/30 19:32:59 INFO fs.TestDFSIO: Test exec time
sec: 27.1
```

To test read performance

$ yarn jar /usr/local/hadoop/share/hadoop/mapreduce/hadoop-mapreduce-client-jobclient-2.7.0.jar TestDFSIO -read -nrFiles 10 -fileSize 100

```
17/04/30 19:43:50 WARN hdfs.DFSClient:
DFSInputStream closed already
17/04/30 19:43:50 INFO fs.TestDFSIO: ----- TestDFSIO
----- : read
17/04/30 19:43:50 INFO fs.TestDFSIO: Date & time:
Sun Apr 30 19:43:50 IST 2017
17/04/30 19:43:50 INFO fs.TestDFSIO: Number of
files: 10
17/04/30 19:43:50 INFO fs.TestDFSIO: Total MBytes
processed: 1000.0
17/04/30 19:43:50 INFO fs.TestDFSIO: Average IO rate
mb/sec: 428.6703186035156
17/04/30 19:43:50 INFO fs.TestDFSIO: IO rate std
deviation: 104.85671091237035
```

To clean the files generated by these benchmarks

$ yarn jar /usr/local/hadoop/share/hadoop/mapreduce/hadoop-mapre-duce-client-jobclient-2.7.0.jar TestDFSIO -clean

```
17/04/30 19:45:01 INFO fs.TestDFSIO: TestDFSIO.1.8
17/04/30 19:45:01 INFO fs.TestDFSIO: nrFiles = 1
17/04/30 19:45:01 INFO fs.TestDFSIO: nrBytes (MB) = 1.0
17/04/30 19:45:01 INFO fs.TestDFSIO: bufferSize =
1000000
17/04/30 19:45:01 INFO fs.TestDFSIO: baseDir = /
benchmarks/TestDFSIO
17/04/30 19:45:02 WARN util.NativeCodeLoader: Unable
to load native-hadoop library for your platform...
using builtin-java classes where
17/04/30 19:45:02 INFO fs.TestDFSIO: Cleaning up
test files
```

6.11 YARN SCHEDULERS

A cluster is shared among many different big data processing tools and frameworks by using YARN. Jobs from all these frameworks are submitted regularly. Resources are a limited commodity in the busy cluster and hundreds of jobs are submitted now and then. Moreover, some specific amount of resources is reserved for users and departments using predefined policies. Therefore, scheduling jobs onto resources available in the cluster preserving constraints becomes a NP Hard problem and there is no better optimal policy. The YARN provides a set of schedulers and configurable policies. The YARN supports three different schedulers: First-In-First-Out (FIFO), capacity, and fair. Goto RM_IP:8088/cluster/scheduler in browser to explore scheduler and its policies.

YARN, by default, comes with the capacity scheduler. However, one can change the scheduler using properties as these schedulers are pluggable. This means that users can tweak the source code of those schedulers or develop new scheduler and include in the YARN. To change the scheduler, you have to include the respective scheduler property in the yarn-site.xml configuration file and restart yarn services. The YARN schedulers are typically evaluated based on the different quality of services such as job latency, throughput, resource utilization, number of data-local execution and non-local execution, fairness among jobs, deadline, etc.

FIFO scheduler

There are no configuration policies for FIFO as it schedules jobs/applications as it arrives. Every job acquires the whole cluster in its turn and does not share with other jobs. However, if there is a small job in the queue, it is assigned with high priority as it should not wait for long. Priorities are VERY_HIGH, HIGH, NORMAL, LOW, and VERY_LOW. However, if the longer job is already taken up for execution, there is no concept of preemption in FIFO. So, priorities do not support preemption. To set FIFO scheduler for YARN, first, stop yarn services, then edit yarn-site.xml file to include the following property, and restart yarn service.

```
<property>
    <name>yarn.resourcemanager.scheduler.class</name>
    <value>org.apache.hadoop.yarn.server.
    resourcemanager.scheduler.fifo.FifoScheduler</
    value>
</property>
```

Capacity scheduler

Capacity is the default scheduler in the YARN and was designed by Yahoo to support multi-user and can execute more than one job simultaneously by sharing the cluster resources. A portion of available resources in a cluster is reserved for each job. Suppose, if there is only one job running in the cluster, it can make use of the entire cluster resources. Once the second job is submitted, it claims its reserved portion of resources from the first job. So, reserved resources for a second job is forcefully (preemption) taken from the first job and given to the second job. This provides elasticity in a cost-effective manner and preserves jobs reservation.

This fixed capacity is guaranteed all the time based upon software-level agreement. Once a user gets some amount of resources, they can be shared further among the jobs of that user in a hierarchical fashion. Each reserved portion of resources employs a queue to schedule jobs. Within a queue, jobs are scheduled using FIFO scheduler. Each queue may occupy the resources of another queue upon availability. However, we can set a threshold that a queue should not exceed consuming other queue resources. The capacity scheduler can be set in the yarn-site.xml file by including the following property and restart the yarn service to take effect.

```
<property>
    <name>yarn.resourcemanager.scheduler.class</name>
    <value>org.apache.hadoop.yarn.server.
    resourcemanager.scheduler.capacity.
    CapacityScheduler</value>
</property>
```

Capacity scheduler queue configuration is specified in /etc/hadoop/ capacity-scheduler. xml file. It has a predefined queue called *root*. All other queues are children of the root queue. Here is an example with three child-queues a, b and c and some sub-queues for a and b:

```
<property>
    <name>yarn.scheduler.capacity.root.queues</name>
    <value>a,b,c</value>
</property>
<property>
    <name>yarn.scheduler.capacity.root.a.queues</name>
    <value>a1,a2</value>
</property>
<property>
    <name>yarn.scheduler.capacity.root.b.queues</name>
    <value>b1,b2,b3</value>
</property>
```

Fair scheduler

Fair scheduler was introduced by Facebook and designed for multi-user/multi-job environment. It equally (fair) divides cluster resources among multiple jobs unlike capacity scheduler having fixed resources for each job. However, similar to capacity scheduler, if only one job is running, it gets all the resources from the cluster. When the second job is submitted, it preempts its fair share from the first job. Preemption is enabled by setting yarn.scheduler.fair. preemption to true. Fair sharing allows short jobs to finish in a reasonable time. Fair sharing can also work with job priorities, which are used as weights to determine the fraction of total resources that each job should get. Edit yarn-site.xml to set fair scheduler:

```
<property>
    <name>yarn.resourcemanager.scheduler.class</name>
    <value>org.apache.hadoop.yarn.server.
    resourcemanager.scheduler.fair.FairScheduler</
    value>
</property>
```

Delay scheduling

The YARN schedulers typically attempt data locality to the maximum for map tasks. If there is no possibility of data locality, the particular data block from busy node is copied to some other node in the cluster to execute map task. This is called non-local execution. However, there may be a chance of data local execution if the particular task waits for few seconds for the completion of already running tasks. The waiting time of a task should not be higher than the time taken to copy data block and executing tasks remotely. This feature is called delay scheduling and supported by both capacity and fair scheduler. Delay scheduling can be configured with capacity scheduler using yarn.scheduler.capacity.node-locality-delay property and fair scheduler using yarn.scheduler.fair.locality.threshold.node (default 0.5) property.

Dominant Resource Fairness (DRF)

If there is single resource-based scheduling, say only memory, scheduling decision is straightforward depending upon the memory requirement of each job. If scheduling is based on multiple resources, it becomes a difficult task to schedule. For instance, one complex scenario is to share resources among jobs, in which one job requires much CPU with little memory, and the other job requires little CPU and more memory. The YARN schedulers address this problem by taking a dominant resource of each job concerning the amount of cluster resources. This is called the DRF.

Consider a cluster with 100 CPUs and 1 TB memory. Application A requests containers of (2 CPUs, 300 GB), and application B requests containers of (6 CPUs, 100 GB). A's request is (2%, 3%) of the cluster, and B's request is (6%, 1%) of the cluster. So, for application A, memory is dominant since its proportion (3%) is larger than CPU (2%). As B's container request is twice as big in the dominant resource (6% versus 3%), B will get containers under fair sharing. DRF is not enabled by default. The capacity scheduler can be configured to use DRF by setting yarn.scheduler.capacity. resource-calculator to `org.apache.hadoop.yarn.util.resource.` `DominantResourceCalculator` in capacityscheduler.xml.

6.12 COMPRESSION

Data blocks and output of map/reduce tasks can be compressed to minimize the size before storing in HDFS or transferring over the local network. This will minimize the disk/network IO latency for map/reduce tasks and the number of map tasks. Consider a 20 GB file with block size 128 MB. If there is no compression, 160 blocks are created, which results in 160 map tasks. There are two types of compression: splittable and non-splittable.

Splittable compression

This allows compressed files to get divided into blocks. After compression, if 20 GB file is turned to be 2 GB, then there will be only 16 blocks. Thus, the number of map tasks is minimized. Splittable compression performs at a lower rate due to the overheads, because it has to identify the physical boundaries of blocks. However, it is highly suitable for the MR jobs.

Non-splittable compression

Once a file is compressed, it cannot be divided into blocks to store onto HDFS. This is called non-splittable compression. So, only one map task is launched to process the entire compressed file regardless of its size. It causes huge non-local execution leading to increase in job latency. Compression schemes differ in terms of the size of output, the number of map tasks, and data locality. However, the output of these compression algorithms is stored in the same way. The compression algorithm is denoted by "codec" (compression and decompression). There are already predefined compression algorithms: deflate, gzip, bzip2, lzo, lz4, snappy, form which only bzip2 is splittable, the rest are non-splittable. We can compress both map and reduce task output by configuring properties or specifying in driver function. To compress map task output, use the following configuration properties in mapred-site.xml file:

```
<property>
    <name>mapreduce.map.output.compress</name>
    <value>true</value>
</property>
<property>
    <name>mapreduce.map.output.compress.codec</name>
    <value>org.apache.hadoop.io.compress.GzipCodec</
    value>
</property>
```

When you include properties in configuration files, you will have to restart the respective service to take effect. This change is permanent until you modify it. Instead, you can specify these properties in driver function or pass as run-time arguments while launching a job.

```
Configuration conf = new Configuration();
conf.setBoolean(Job.MAP_OUTPUT_COMPRESS, true);
conf.setClass(Job.MAP_OUTPUT_COMPRESS_CODEC,
GzipCodec.class,
CompressionCodec.class);
Job job = Job.getInstance(conf,"wordcount");
```

To compress the job output (reduce task output) using configuration properties:

```
<property>
    <name>mapreduce.output.fileoutputformat.
    compress</name>
    <value>true</value>
</property>
<property>
    <name>mapreduce.output.fileoutputformat.compress.
    codec</name>
    <value>org.apache.hadoop.io.compress.Bzip2Codec</
    value>
</property>
<property>
```

```
<name>mapreduce.output.fileoutputformat.type</
name>
<value>RECORD</value>
</property>
```

To compress the job output using driver function:

```
FileOutputFormat.setCompressOutput(job,true);
FileOutputFormat.setOutputCompressorClass(job,GzipC
odec.class);
```

Compression is done either block-level or record-level. Block-level compression is much better and faster. Compression algorithms trade-off space and time. Faster algorithms provide smaller space savings and slower algorithms need more space in memory. There are 9 options in this regard. -1 means to optimize speed and -9 means to optimize space. Gzip sits in the middle of space-time trade off.

Compression algorithms are available in native (other than Java) implementation, but deflate, gzip, and bzip2 algorithms are java implementation. In some scenario, native implementation is much faster than java implementation. However, to use native implementation of the algorithm, Hadoop source code must be compiled together with native algorithm codes. It is good to compress after splitting the file into blocks. However, the compressed blocks should be approximately the HDFS block size. For a large input file, you should not use a compression algorithm that is not splittable, which will largely increase non-local execution.

6.13   DISTRIBUTED CACHE

A distributed cache is a portion of memory allocated to load simple read-only files/archives at run time in NMs apart from map/reduce task input. Its size can be up to a few MBs. These read-only files are passed as command line arguments (or generic options such as -files, -archives, -jars) or specified in driver function as shown in Table 6.5. When a job is launched, these files will be pushed into HDFS. Later, YarnChild will load them into its local directory before starting map/reduce task. This is highly useful for applications like join operation with smaller data sets. We use setup()in map/reduce task to open such files in the cache to read. For instance, consider weather analysis application that needs to find the maximum temperature recorded in each city. Temperature is recorded along with city code, not with the city name.

There is a file "index" containing the name of each city with city code. So, the application would require every reduce task to read the index file to find the city name for the corresponding city code. Steps are given below to implement a distributed cache.

- Prepare dataset to pass into a distributed cache.
- Write a parser class for the dataset (optional).
- Launch a job with -files option.
- Use GenericOptionsParser to get the file location.
- Build HashMap in map/reduce task to look up using keys.

There are some properties to customize distributed cache before passing input files. These properties are included in yarn-site.xml file. Some important configuration properties are given below:

yarn.nodemanager.local-cache.max-files-per-directory	- 8192 (in MBs)
yarn.nodemanager.local-dirs	- ${hadoop.tmp.dir}/nm-local-dir
yarn.nodemanager.localizer.cache.cleanup.interval-ms	- 600000
yarn.nodemanager.localizer.cache.target-size-mb	- 10240 (in MBs)
yarn.nodemanager.localizer.client.thread-count	- 5
yarn.nodemanager.localizer.fetch.thread-count	- 4

There are different ways to pass input files to distributed cache as shown in Table 6.5 when the job is launched. You can pass files to distributed cache using generic options (-files, -archives, -jar) or hard code in driver function using APIs.

TABLE 6.5   Passing Files to Distributed Cache

Job API	Generic options	Description
addCacheFile(URI uri) setCacheFiles(URI[] files)	-files file1, file2	to add files into distributed cache.
addCacheArchive(URI uri) setCacheArchives(URI[] uri)	-archives archive1, archive2	to add archives into distributed cache. It is archived once copied into containers.
addFiletoClassPath(Path file)	-libjars jar1, jar2	to add jars into distributed cache for classpath
addArhciveToClassPath(Path archive)	-	to add files into distributed cache as MR tasks classpath

6.14 JOINING DATASET

Combining attributes/fields of different datasets is called join. It does not increase the number of rows/records but increases the number of columns in the resulting table. Joins are a bit expensive in MR as it requires two datasets to traverse. If one dataset is large and another one is small in a join operation, the smaller dataset can be passed to map/reduce task via distributed cache. If both input datasets are large, then MR takes more time to finish join operation. However, joins are relatively smooth with pig, hive, cascading, spark when compared to MR.

Input dataset1

Country code	Country name
001	USA
002	England
003	China

Input dataset2

Country code	Population	Year
001	321.4 million	2015
002	53.01 million	2011
003	1.311 billion	2015
001	350 million	2017

Output dataset

Country code	Country name	Population	Year
001	USA	321.4 million	2015
002	England	53.01 million	2011
003	China	1.311 billion	2015
001	USA	350 million	2017

Map-side join

If the join is performed in map task, it is called map-side join. Map task should use join key as map output key and partition map output based on the join key. For the above example, join key is country code. So, the records with similar join key are brought to respective reduce task. A map-side join can be used to join the output of several jobs, but it is less efficient because both datasets have to go through MR shuffle.

Reduce-side join

If join is performed in reduce task, it is called reduce-side join. Reduce-side join is more general than map-side join. To suit reduce side join for your application, you may have to implement custom writable comparable, custom partitioner, and custom group comparator. Outer join and inner join can be applied in both map and reduce task. Moreover, join operations may require a chain of map/reduce tasks.

6.15 SERIALIZATION

Processes communicate with one another. For example, reduce tasks collect output from all map tasks. Inter-process communication is the backbone of a distributed system, and communication among Hadoop daemons is done by RPC. Data transferred across the network must be serialized/deserialized properly. RPC performs serialization and de-serialization to send data back and forth. Data serialization is a process of packing structured objects in a standard format and writing them onto byte stream to either store or exchange across other heterogeneous applications in a compact way. Deserialization is the reverse process of turning byte stream back into a series of structured objects. Serialized data formats are represented in XML/JSON/BSON.

Some of the data serialization tools are Avro, Apache Thrift, Google Protobuff. Serialization and deserialization are done using the Writable interface. Sometimes the map task is written in Java and reduce task is written in python. Now, data exchange among these two different languages is not compatible as their datatype size is different. So, we use Avro, Thrift, Protobuff, etc. to achieve this cross-language compatibility. A serialization tool must provide the following features:

- Compact: data transmitted over the network must be small, the smaller the data, the higher the network efficiency.
- Fast: serialization and de-serialization should be quicker.
- Extensible: protocol change over time, so it should be feasible to add more ideas like including more arguments later.
- Interoperable: cross language compatibility.

Why cannot Hadoop use java serialization? It has the following shortcomings:

- Java serialization writes the class name, meta-data of the object along with data on the stream. So, the size of the serialized data is huge and takes more time to de-serialize.

- Sorting and random access are difficult to perform.
- The de-serialization process in java creates a new instance of the object, while Hadoop needs to reuse objects to minimize computation.
- RMI gives much control over serialization. However, it did not satisfy the serialization features.

Therefore, Hadoop has its serializable format with Writable and WritableComparable interfaces (available in org.apache.hadoop.io.serializer package). However, these interfaces are language dependent. So, it is better to use Protobuff or Avro for serialization instead of the Writable interface. If integer datatype itself supports serialization, then it is IntWritable. Similarly, Table 2.4 summarizes all MR datatypes that support serialization. All the Writable data types are mutable (content modifiable) except NullWritable. Most of the methods in java String work with characters while methods in Text work on Unicode bytes. The string type is immutable while Text is mutable. UTF-8 format for Text helps to interoperate with other languages easily and is equivalent to String in Java. However, to manipulate the text efficiently, String has more featured methods compared to Text. So, it is recommended to convert the Text object of MR to String to perform various text manipulations. Some special cases are:

NullWritable – It is a special type and does not output anything as part of the key/value.

ObjectWritable – It is a general-purpose wrapper for java primitive types, String, enum, Writable, null, arrays of any data types, etc. Hadoop RPC uses them for serialization. It is helpful when a record comprises a mix of different data types.

GenericWritable – It is used when you know the type of data, so you need not do any type of casting later.

Writable collections – It includes ArrayWritable, ArrayPrimitiveWritable, TwoDArray Writable, MapWritable, SortedMapWritable, EnumSetWritable, etc.

Why custom writable?

Hadoop comes with a useful set of writable MR datatypes that serve most purposes. Still, Hadoop provides an option to devise custom datatypes that give the full control over the binary representation and the sort order. However, rather than developing custom datatypes, it is better trying another framework like Avro that allows you to define custom data types declaratively.

Writable and its importance in Hadoop

Writable is an interface in Hadoop and acts as a wrapper class for almost all the primitive data types of Java. That is how int of java has become IntWritable in Hadoop and String of Java has become Text in Hadoop. Writable is used for creating serialized data types in Hadoop. Serialization is not the only concern of Writable interface; it also has to perform, compare, and sort operation in Hadoop. As explained above, if a key is taken as IntWritable, by default it has a comparable feature because of RawComparator acting on that variable. Therefore, it gives compare flexibility during sort in the execution sequence.

What happens if WritableComparable is not present?

If we create custom datatype with Writable (not WritableComparable), it will not have the capability to get compared with other datatypes. There is no compulsion that custom datatypes need to be WritableComparable until and unless if it is a key. Because, values do not need to be compared with each other as keys (remember merge, sort, and group steps in MR execution sequence?). If the custom data type is a key, it should implement WritableComparable or else the keys cannot be compared during MR execution sequence.

6.16   GLOBAL (TOTAL) SORTING

Sorting is done in both map and reduce functions. In the map, intermediate map outputs are partitioned and sorted before spilling into the disk. After too many spills, all the respective partitions are merged into one and then sent to the appropriate reducer. Before reduce function starts, again merge and sort operation are done. Sorting at the map side minimizes huge effort at reducer side. Now, the question is, how output files from reduce tasks of a job are sorted globally?

Partial sort

It means that when you use 10 reduce tasks, you will get 10 output files sorted locally, but not globally. Merging all 10 output files will not produce the global sorted version.

Total sort

How can you produce 10 output files sorted globally? There are different ways:

1. Use a single reducer. However, it does not achieve parallelism and takes more time.
2. Do not set reducer at all. But, you will get only map output as the sorted version by using a default reducer. So, the same key gets redundant values.
3. Multiple output file: For each reduce function input key, a file is created. Finally, you have to use getmerge command to combine all the output files into one big sorted file.
4. If we use multiple reduce tasks, one way of doing global sorting is to have a custom partitioner with range partitioning. For this to work, you have to know the range of all map output key. Example: If there are three reduce tasks for wordcount job, you can do range partitioning as follows:
   - the word that starts with a to g to go to reducer1
   - the word that starts with h to s to go to reducer2
   - the word that starts with s to z to go to reducer3

   The output of each reduce task is sorted and using getmerge command you will get globally sorted output. However, it may lead to data skewness among reduce tasks. For instance, if there are one million words in a file, 80% goes to reduce task1, 15% goes to reduce task2, and the rest goes to reduce task3. This increases job latency. Therefore, you have to understand the key range of the whole dataset to distribute the values. It is the programer responsibility to be aware of the key range. However, it is possible to get fairly even set of partitions by sampling the key space. Hadoop provides a few default samplers (random sampler).
5. Once all MR job completed, you can run pig/hive queries to get global sorting.

Comparator

Sorting happens using a sorting comparator. Grouping is done using grouping comparator. By default compareTo() function is used to sort. You can customize sorting and grouping by overriding compareTo(). To work on sorting effectively, you need to have an idea about partitioner, comparator, multiple output formats, etc.

RawComparator

compareTo() from java.lang.Comparable interface performs object-level comparison and requires serialization and de-serialization of the underlying objects. It impacts performance while dealing with huge data. To overcome this, the MR framework provides an interface called RawComparator to implement compareTo() behavior. It performs byte-level comparison without serialization and de-serialization as it extends BinaryComparable class. In MRv2, shuffle, and sort also are pluggable. Therefore, you can incorporate your idea.

Secondary Sort

A composite key contains more than one field. Sorting using a field in the composite key is called a secondary key. Example: we have weather dataset and need to find the maximum temperature each year.

> Map task – map function reads a record, then outputs year (key) and temperature (value) for each record from the dataset.
> Reduce task – receives 1000s of temperature values for each key (year), and iterates all temperatures to find the maximum one.

Here, a composite key can be formed taking year and temperature as part of the key for map task output.

> composite key → (field1, field2) → (year, temperature)
> map task → (composite key, value) → partitioner (field1) → shuffle
> merge → sort (field2) → group → reduce function

User can define partitioning, sorting, and grouping based on any field in the composite key. In our example, we partition based on field1 (year), and sort based on field2 (temperature). Now, it is pretty easy to pick the maximum temperature without iterating millions of temperature values in reduce function.

6.17   FILTERING

If there is a directory with 1000s of files, it is possible to filter out only the required files to pass as input to MR job. In MR application development, filtering can be done

- before invoking map/reduce task (using FileSystem or FileInputFormat)
- in map task
- in reduce task

The MR applications should be designed in a way that data is filtered as early as possible. This can be achieved using the following ways as given below:

- using **org.apache.hadoop.fs** package classes in driver function itself.
- using setup(): filtering conditions can be passed externally as command line arguments. In this case, we need to use setup() in the map/reduce task.
- FileInputFormat has filters to eliminate paths that are not required to process.
- Configuration API.

Most of the file filtering logic should be implemented as part of the map task. Path filtering can be done only on names and pattern on the name (not on properties such as file creation time, file modified time). There is an option globStatus in FileSystem class to ease filtering.

Glob	name	matches
*	asterisk	matches zero or more characters
?	question mark	matches a single character
[ab]	character class	matches single character in the set {a,b}
[*ab]	negated character	matches a character that is not in the set {a,b}
[a-b]	character range	matches a single character in the closed range [a,b]

Passing a directory as input to the MR job should not contain any subdirectories, else exception will be thrown. We have to pass the path of each subdirectory separately. File input format uses a default filter that excludes hidden files beginning with special characters (. or _). User also can define setInputPathFilter().

6.18   CONVERTING HADOOP SOURCE CODE TO BINARY CODE

While running commands in CLI, you would have observed a warning in the first line followed by the result. Example:

```
$ hdfs dfs -ls /
```

```
WARN: util.NativeCodeLoader: Unable to load native-hadoop
library for your platform... using built in-java classes
where applicable/ Found 2 items
```

```
-rw-r--r-- 3 itadmin supergroup 693 2017-05-08 14:34
/cards.txt
-rw-r--r-- 3 itadmin supergroup 14 2017-05-08 15:43
/input.txt
The reason for this warning is, we downloaded Hadoop
binary file /usr/local/hadoop/lib/ native/libhadoop.
so.1.0.0 which was already compiled on a different
platform. To avoid such warnings, including the
following in Hadoop environment file.
```

$ sudo vi /usr/local/hadoop/etc/hadoop/hadoop-env.sh

```
export HADOOP_OPTS="$HADOOP_OPTS -Djava.library.
path=$HADOOP_HOME/lib/native"
```

$ hdfs dfs -ls /   // no warnings are displayed

However, to work with compression, security, and cluster management, you need to explicitly convert Hadoop opensource code into binaries that are compatible with your current platform. Therefore, we are going to see how the source code version of Hadoop can be converted to our native (local machine) binary on 64-bit Ubuntu18 platform. The steps are:

1. Download and install the latest version of java and set path (see Section 3.1).
2. Install maven and set path. Maven is used to compile and build source code with required dependent jars. If we do not use maven, then we have to use Eclipse. Let us continue with maven.

    $ wget http://mirrors.estointernet.in/apache/maven/maven-3/3.6.1/binaries/apache-maven-3.6.1-bin.tar.gz
    $ tar -zxvf apache-maven-3.6.1-bin.tar.gz
    $ sudo cp -r apache-maven-3.6.1 /usr/local/maven
    $ vi .bashrc

    ```
    export MAVEN_HOME=/usr/local/maven
    export PATH=$PATH:$MAVEN_HOME/bin
    ```
    $ source .bashrc
    $ $PATH
    $ mvn -version   // should display version 3.6.1
    $ mvn -help   // should display all commands

3. Download source code of Hadoop project

Opensource code size is very less because no dependencies (supporting jars) have been included. If you compile the source code using Maven, it automatically downloads all supporting jars. So, Hadoop size will get fat. Go to the following link to download Hadoop source code. https://archive.apache.org/dist/hadoop/core/

$ wget https://archive.apache.org/dist/hadoop/core/hadoop-2.7.0/hadoop-2.7.0-src.tar.gz

$ tar -zxvf hadoop-2.7.0-src.tar.gz

$ ls hadoop-2.7.0-src   // you will see all folders having source code and pom.xml

```
BUILDING.txt   hadoop-hdfs-project   hadoop-tools
dev-support    hadoop-mapreduce-project
               hadoop-yarn-project
hadoop-assemblies  hadoop-maven-plugins
               LICENSE.txt
hadoop-client  hadoop-minicluster
               NOTICE.txt
hadoop-common-project       hadoop-project      pom.xml
hadoop-dist    hadoop-project-dist   README.txt
```

$ vi hadoop-2.7.0-src/BUILDING.txt // contains information about
                                    source code compilation.

Use checknative command to check whether any native libraries are available for the current platform or not. You will get errors, warnings for the package not included in the binary version of the current Hadoop file.

$ hadoop checknative -a

```
17/03/30 15:05:59 WARN util.NativeCodeLoader: Unable
to load native-hadoop library for your platform...
using builtin-java classes where applicable. Native
library checking:
    hadoop:      false
    zlib:        false
    snappy:      false
    lz4:         false
    bzip2:       false
    openssl:     false
```

```
17/03/30 15:06:00 INFO util.ExitUtil: Exiting with status 1
```

4. Make passwordless connection
   $ ssh-keygen or ssh-keygen -t rsa
   $ cat ~/.ssh/id_rsa.pub >> ~/.ssh/authorized_keys
   $ chmod 0700 ~/.ssh/authorized_keys

5. Install the following tools to convert source code to target platforms.
   $ sudo apt-get -y install build-essential autoconf automake libtool cmake zlib1g-dev pkg-config libssl-dev

   Download and install the latest protocol buffer for interprocess communication among processes.

   $ wget https://github.com/protocolbuffers/protobuf/releases/download/v2.5.0/protobuf-2.5.0.tar.gz

   $ tar -zxvf protobuf-2.5.0.tar.gz
   $ cd protobuf-2.5.0/
   $ ./configure
   $ make
   $ sudo make install
   $ protoc --version
   $ sudo reboot   // reboot the server to effect protoc

6. Optional packages to install
   Snappy compression: $ sudo apt-get install snappy libsnappy-dev
   Bzip2: $ sudo apt-get install bzip2 libbz2-dev
   Jansson (C Library for JSON): $ sudo apt-get install libjansson-dev
   Linux FUSE: $ sudo apt-get install fuse libfuse-dev

7. Convert source code to target platform by downloading required jars.
   $ cd hadoop-2.7.0-src
   $ mvn package -Pdist,native -DskipTests -Dtar
   At the end of execution, you will see a list of libraries downloaded successfully.

```
[INFO] Reactor Summary:
[INFO] Apache Hadoop Main......SUCCESS [ 0.648 s]
[INFO] Apache Hadoop Project POM...SUCCESS [ 0.617 s]
[INFO] Apache Hadoop Annotations...SUCCESS [ 1.143 s]
[INFO] Apache Hadoop Assemblies...SUCCESS [ 0.073 s]
[INFO] Apache Hadoop Project Dist POM.......SUCCESS
[ 1.210 s]
[INFO] Apache Hadoop Maven Plugins...SUCCESS [ 1.343 s]
```

```
[INFO] Apache Hadoop MiniKDC...SUCCESS [ 1.069 s]
[INFO] Apache Hadoop Auth.......SUCCESS [ 1.395 s]
[INFO] Apache Hadoop Auth Examples...SUCCESS [ 1.634 s]
```

There should not be any failure cases. All should be successful. Rerun the same command if is unsuccessful.

8. Check the jars and native libraries created in the location hadoop-2.7.0-src/hadoop-dist/ target/. To see the native libraries generated,

$ ls hadoop-2.7.0-src/hadoop-dist/target/hadoop-2.7.0/lib/native/

```
libhadoop.a libhadoop.so libhadooputils.a libhdfs.so
libhadooppipes.a libhadoop.so.1.0.0 libhdfs.a
libhdfs.so.0.0.0
```

9. Launch newly generated Hadoop files.

Option 1: If you have already Hadoop single node/multi-node cluster up and running, just replace its native libraries with these newly generated libraries.

$ sudo cp -r hadoop-2.7.0-src/hadoop-dist/target/hadoop-2.7.0/lib/ native/ /usr/local/ hadoop/lib/native/

Option 2: You can use hadoop-2.7.0-src/hadoop-dist/target/hadoop-2.7.0 file itself to set up single node or multi-node cluster.

For both the options you have to include native library location in hadoop-env.sh

$ sudo vi /usr/local/hadoop/etc/hadoop/hadoop-env.sh

```
export HADOOP_OPTS="$HADOOP_OPTS -Djava.library.
path=$HADOOP_HOME/lib/native"
```

$ hadoop checknative -a   // you will see all native libraries and all are true.

Native library checking:

```
hadoop: true /usr/local/hadoop/lib/native/libhadoop.
so.1.0.0
zlib: true /lib/x86_64-linux-gnu/libz.so.1
snappy: true /usr/lib/x86_64-linux-gnu/libsnappy.
so.1
lz4: true revision:99
bzip2: false
openssl: true /usr/lib/x86_64-linux-gnu/libcrypto.so
```

You can install other packages mentioned in the optional package. In the above output, the bzip2 compressor is false. If you want to install bzip2 and include into binaries:

$ sudo apt-get install bzip2 libbz2-dev
$ mvn package -Pdist,native -DskipTests -Dtar // now, it is faster as
already all jars downloaded
$ hdfs dfs -ls /

It does not show warning anymore. You can create binaries for Windows OS also. However, it takes a different set of commands and requires some library packages to be installed.

6.19   HDFS FEDERATION

The primary responsibilities of NN are managing and maintaining:

- namespace and meta-data in memory, co-ordinating DN operations.
- cluster membership by registering/deregistering DNs via periodic HB from DNs.
- block locations and block under/over replications.

The number of data blocks stored in HDFS depends on the amount of memory available in the NN. Meta-data for objects (directory/file/block) takes 150 bytes in memory.

Two objects = 1 file + 1 block = (1 * 150) bytes + (1 * 150) bytes = 300 bytes 1 billion objects = approximately 14 GB

As the number of blocks increases, it is difficult to manage huge meta-data in memory. Moreover, the NN takes some time to search for data from the huge meta-data. Therefore, Hadoop 2.x allows to set up multiple NNs (also called federation) as shown in Figure 6.6 to share loads of single NN. The benefits of using HDFS federation are:

- HDFS scalability – cluster storage capacity scales horizontally as each NN can support several 1000s of slaves.
- Performance – multiple NN in cluster serves file system read/write operations simultaneously. Therefore, the load is balanced, and the throughput increases linearly.
- It offers isolation in a multi-user environment. For instance, if there are several departments, each department data can be maintained by individual NN.

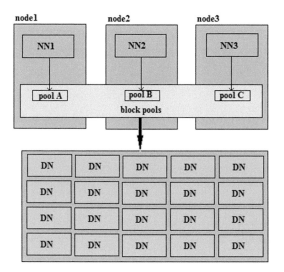

FIGURE 6.6 HDFS federation.

There are some terminologies to understand before diving into NN federation and debugging.

Block pool

Each NN has its namespace and hosts a pool of blocks, as shown in Figure 6.6. Every DN store blocks of different pools. One NN generates blockID for its blocks regardless of other NN. The NNs do not coordinate themselves for generating global blockID. The NN does not share its namespace with other NN. HDFS federation supports up to 10,000 nodes in a cluster.

Namespace volume

The amount of data a NN can handle is called namespace volume. It is not the same for each NN. The NNs do not communicate with one another, and the failure of one NN does not affect the availability of the namespace managed by other NNs. Each DN registers with all NNs and sends BR and HB to all NNs. One NN failure does not prevent DNs from serving other NNs in the cluster. However, blocks that belong to failed NN cannot be accessed. When the NN or namespace is deleted, the corresponding block pool at DN is deleted. To delete a block pool from a DN,

```
$ hdfs dfsadmin -deleteBlockPool <DN-host:port> blockpoolId
```

ClusterID

It is an identifier used to identify the DNs in the cluster. Upon NN format, it is either auto-generated or provided manually.

ViewFS

Clients can get a unified view of multiple namespace via an API called ViewFS. So, the client will see only one logically unified namespace in the WUI despite the existence of several NN namespaces.

6.19.1   HDFS FEDERATION DEPLOYMENT

Consider five nodes (PM/VM) as given in Table 6.6. Follow the requirements given in Section 3.1 for all the nodes in the cluster before beginning installation. Refer Figure 3.5 and Figure 5.3 to visualize how Hadoop 2.x multi-node is set up on a cluster of PMs and a cluster of VMs. Refer Tables 3.2 and 3.3 to know the possible cluster configuration for Hadoop.

TABLE 6.6   HDFS NN Federation

PM/VM	Daemon	IP	Hostname	Domain name
node1	NN1	10.100.55.92	ubuntu1	node1
node2	RM	10.100.55.93	ubuntu2	node2
node3	Slave1 (DN+NM)	10.100.55.94	ubuntu3	node3
node4	Slave2 (DN+NM)	10.100.55.95	ubuntu4	node4
node5	NN2	10.100.55.96	ubuntu5	node5

Values such as IP, hostname, and domain name in Table 6.6 are based on my cluster. You have to fill this table according to your cluster environment. We are going to configure NN federation with node1 for NN1 and node5 for NN2. Username must be the same in all the nodes in the cluster. I assume username for all the nodes "**itadmin.**" Moreover, the hostname must be different in all the nodes. Refer Section 6.2 for some basic requirements to achieve the multi-node cluster. Try multi-node installation before starting HDFS federation installation.

6.19.1.1   *OS TYPE AND JAVA VERSION IN ALL PHYSICAL/VIRTUAL NODES*

1. OS type must be same in all nodes (either 32-bit OS or 64-bit OS, but not mixed up). Use the following command to check OS type. If it is not the same, it is better to install the right version.
    $ uname -a
    Result for 32-bit Ubuntu will be as below:

    ```
    Linux discworld 2.6.38-8-generic #42-Ubuntu SMP
    Mon Apr 11 03:31:50 UTC 2011 i686 i686 i386 GNU/
    Linux
    ```

    Result for 64-bit Ubuntu will be as below:

    ```
    Linux ubuntu 4.4.0-104-generic #127-Ubuntu SMP Mon
    Dec 11 12:16:42 UTC 2017 x86_64 x86_64 x86_64 GNU/
    Linux
    ```

    Alternatively, you can use the following command
    $ uname -i

    ```
    x86_64
    ```

2. Install Java in all the nodes with the same version (OpenJDK/Oracle JDK) and architecture (32-bit/64-bit). Refer Section 3.1 to install Java. To check java version and architecture,
    $ java -version

    ```
    openjdk version "1.8.0_131"
    OpenJDK Runtime Environment (build
    1.8.0_131-8u131-b11-2ubuntu1.16.04.3-b11)
    OpenJDK 64-Bit Server VM (build 25.131-b11, mixed
    mode)
    ```

    $ javac -version

    ```
    javac 1.8.0_131
    ```

    or go to /usr/lib/jvm
    $ ls /usr/lib/jvm
    Result for 32-bit java will be as below:

    ```
    java-1.8.0-openjdk-i386 java-8-openjdk-i386
    ```

    Result for 64-bit Ubuntu:

    ```
    java-1.8.0-openjdk-amd64 java-8-openjdk-amd64
    ```

Caution:

1. Username must be the same in all the nodes. I have set **"itadmin"** as the username for all the nodes in my cluster. So, create a common username in all nodes in the cluster using the following command.

   $ sudo adduser itadmin   // leave empty for all the fields
   $ sudo adduser itadmin sudo

   Now, username itadmin is created. Log into itadmin and do the rest.

2. Another important restriction is, the hostname of the nodes must be unique across the cluster; that is, the hostname should not be identical in the cluster. To change the hostname to ubuntu1 in your machine without restarting, use the following commands

   $ sudo vi /etc/hostname

   ```
       ubuntu1
   ```

   $ sudo vi /etc/hosts

   ```
       127.0.0.1    localhost
       127.0.1.1    ubuntu1
   ```

   $ sudo sysctl kernel.hostname=ubuntu1

Do this in all nodes and create unique hostname as given in Table 6.6.

6.19.1.2   INSTALLATION STEPS

Step 1: Set the domain name in all nodes in the cluster

Go to /etc/hosts to set DNS. After setting the domain name, you can use domain names instead of IP address. Comment 127.0.*.1 and enter IP, domain name, and hostname of each node in a line with tab space separator (refer Table 6.6).

   $ sudo vi /etc/hosts

   ```
       #127.0.0.1  localhost
       #127.0.1.1  ubuntu1
       # IP            domain_name         host_name
       10.100.55.92        node1           ubuntu1
       10.100.55.93        node2           ubuntu2
       10.100.55.94        node3           ubuntu3
       10.100.55.95        node4           ubuntu4
       10.100.55.96        node5           ubuntu5
   ```

Step 2: Set passwordless communication

Create an SSH key in NN1 and send to all other nodes. Similarly, create an SSH key in NN2 and send it to all the other nodes.

Generate key in node1

```
$ ssh-keygen or ssh-keygen -t rsa        // leave empty for name, password
$ ls ~/.ssh/
$ cat ~/.ssh/id_rsa.pub >> ~/.ssh/authorized_keys
$ chmod 0700 ~/.ssh/authorized_keys
```

Send keys from node1 to all other nodes in the cluster: the syntax is

```
$ ssh-copy-id -i ~/.ssh/id_rsa.pub username@IP
$ ssh-copy-id -i ~/.ssh/id_rsa.pub itadmin@node2
$ ssh-copy-id -i ~/.ssh/id_rsa.pub itadmin@node3
$ ssh-copy-id -i ~/.ssh/id_rsa.pub itadmin@node4
$ ssh-copy-id -i ~/.ssh/id_rsa.pub itadmin@node5
```

Similarly, generate key in node2 and node5, and send to all other nodes in the cluster.

**Step 3 to Step 6 must be done in all nodes
(masters and slaves) in the cluster**

Step 3: Download Hadoop 2.7.0 and move into hadoop folder

https://archive.apache.org/dist/hadoop/core/
https://archive.apache.org/dist/hadoop/core/hadoop-2.7.0/hadoop-2.7.0.tar.gz

```
$ tar -zxvf hadoop-2.7.0.tar.gz
$ sudo cp -r hadoop-2.7.0 /usr/local/hadoop
```

Step 4: Set path to Hadoop

```
$ vi .bashrc
    export  HADOOP_HOME=/usr/local/hadoop
    export  PATH=$PATH:$HADOOP_HOME/bin
    export  PATH=$PATH:$HADOOP_HOME/sbin
    export  HADOOP_HDFS_HOME=$HADOOP_HOME
    export  HADOOP_MAPRED_HOME=$HADOOP_HOME
    export  HADOOP_COMMON_HOME=$HADOOP_HOME
    export  YARN_HOME=$HADOOP_HOME
    export  HADOOP_CONF_DIR=$HADOOP_HOME/etc/hadoop
```

```
export HADOOP_COMMON_LIB_NATIVE_DIR=$HADOOP_HOME/
lib/native
export HADOOP_OPTS="-Djava.library.
path=$HADOOP_HOME/lib"
```
$ source .bashrc
$ $PATH

Step 5: Configure Java_Home, disable IPv6 in hadoop-env.sh

$ sudo vi /usr/local/hadoop/etc/hadoop/hadoop-env.sh

```
export JAVA_HOME= /usr/lib/jvm/
java-8-openjdk-amd64
export HADOOP_OPTS=-Djava.net.
preferIPv4Stack=true
```
$ hadoop version

Step 6: Edit configuration files

To set up YARN

$ sudo vi /usr/local/hadoop/etc/hadoop/yarn-site.xml

```
<configuration>
    <property>
        <name>yarn.nodemanager.aux-services</name>
        <value>mapreduce_shuffle</value>
    </property>
    <property>
        <name>yarn.nodemanager.aux-services.mapreduce.
        shuffle.class</name> <value>org.apache.hadoop.
        mapred.ShuffleHandler</value>
    </property>
    <property>
        <name>yarn.resourcemanager.hostname</name>
        <value>ubuntu2</value>
    </property>
    <property>
        <name>yarn.resourcemanager.address</name>
        <value>node2:8030</value>
    </property>
```

```
  <property>
    <name>yarn.resourcemanager.resource-tracker.
    address</name>
    <value>node2:8031</value>
  </property>
  <property>
    <name>yarn.resourcemanager.scheduler.address</
    name>
    <value>node2:8032</value>
  </property>
  <property>
    <name>yarn.resourcemanager.admin.address</
    name>
    <value>node2:8033</value>
  </property>
  <property>
    <name>yarn.resourcemanager.webapp.address</
    name>
    <value>node2:8088</value>
  </property>
  <property>
    <name>yarn.nodemanager.disk-health-checker.
    min-healthy-disks</name>
    <value>0.0</value>
  </property>
  <property>
    <name>yarn.nodemanager.disk-health-checker.
    max-disk-utilization-per-disk-percentage</
    name>
    <value>98.5</value>
  </property>
</configuration>
```

To set up MR

$ sudo cp /usr/local/hadoop/etc/hadoop/mapred-site.xml.template /usr/
local/hadoop/etc/hadoop/mapred-site.xml

```
$ sudo vi /usr/local/hadoop/etc/hadoop/mapred-site.xml
    <configuration>
        <property>
            <name>mapreduce.framework.name</name>
            <value>yarn</value>
        </property>
        <property>
            <name>mapreduce.jobhistory.address</name>
            <value>node2:10020</value>
        </property>
        <property>
            <name>mapreduce.jobhistory.webapp.address</
            name>
            <value>node2:19888</value>
        </property>
        <property>
            <name>mapreduce.jobhistory.
            intermediate-done-dir</name>
            <value>/usr/local/hadoop/tmp</value>
        </property>
        <property>
            <name>mapreduce.jobhistory.done-dir</name>
            <value>/usr/local/hadoop/tmp</value>
        </property>
    </configuration>
```

To set up HDFS-related service

```
$ sudo vi /usr/local/hadoop/etc/hadoop/hdfs-site.xml
    <configuration>
        <property>
            <name>dfs.nameservices</name>
            <value>node1,node5</value>
        </property>
        <property>
            <name>dfs.namenode.rpc-address.node1</name>
            <value>node1:10001</value>
        </property>
```

```
    <property>
        <name>dfs.namenode.rpc-address.node5</name>
        <value>node5:10001</value>
    </property>
    <property>
        <name>dfs.permissions</name>
        <value>false</value>
    </property>
</configuration>
```
dfs.permissions – allows any users to create/delete files in HDFS.

To set up slave (DN and NN)

$ sudo vi /usr/local/hadoop/etc/hadoop/slaves
```
    node3
    node4
```

Step 7: To set up NN federation

To be done in node1 for NN1

$ sudo vi /usr/local/hadoop/etc/hadoop/core-site.xml
```
    <configuration>
        <property>
            <name>fs.defaultFS</name>
            <value>hdfs://node1:10001</value>
        </property>
        <property>
            <name>hadoop.tmp.dir</name>
            <value>/usr/local/hadoop/tmp</value>
        </property>
    </configuration>
```

To be done in node5 for NN2

$ sudo vi /usr/local/hadoop/etc/hadoop/core-site.xml
```
    <configuration>
        <property>
            <name>fs.defaultFS</name>
            <value>hdfs://node5:10001</value>
        </property>
```

```
    <property>
        <name>hadoop.tmp.dir</name>
        <value>/usr/local/hadoop/tmp</value>
    </property>
</configuration>
```

Do the following in RM and all slave nodes (node2, node3, and node4)

$ sudo vi /usr/local/hadoop/etc/hadoop/core-site.xml

```
    <configuration>
        <property>
            <name>fs.defaultFS</name>
            <value>viewfs://fedcluster/</value>
        </property>
        <property>
            <name>fs.viewfs.mounttable.fedcluster.link./
            node1</name>
            <value>hdfs://node1:10001</value>
        </property>
        <property>
            <name>fs.viewfs.mounttable.fedcluster.link./
            node5</name>
            <value>hdfs://node5:10001</value>
        </property>
        <property>
            <name>hadoop.tmp.dir</name>
            <value>/usr/local/hadoop/tmp</value>
        </property>
    </configuration>
```

Step 8: Grant ownership and access rights

To be done in all (master and slave) nodes in the cluster

$ sudo chown -R itadmin /usr/local/hadoop
$ sudo chmod -R 777 /usr/local/hadoop

Step 9: Format NN1 and NN2

$ hdfs namenode -format -clusterID fedcluster   // in node1 for NN1
$ hdfs namenode -format -clusterID fedcluster   // in node5 for NN2

Step 10: Start HDFS and YARN

In NN1 (node1) or NN2 (node5)

    $ start-dfs.sh

    $ jps // Check in node1 and node5 for NN service and slave node for DN service

In RM (node2)

    $ start-yarn.sh

    $ mr-jobhistory-daemon.sh start historyserver

    $ jps

NamespaceID, clusterID, blockpoolID: you can verify these in node1 and node5

ClusterID will be same for both NNs. However, blockpoolID and namespaceID will be different.

In node1,

```
$ sudo vi /usr/local/hadoop/tmp/dfs/name/current/VERSION
    namespaceID=909573120
    clusterID=fedcluster
    cTime=0
    storageType=NAME_NODE
    blockpoolI
    D=BP-489835335-10.100.55.92-1493373813213
    layoutVersion=-63
```

In node5,

```
$ sudo vi /usr/local/hadoop/tmp/dfs/name/current/VERSION
    namespaceID=1265114300
    clusterID=fedcluster
    cTime=0
    storageType=NAME_NODE
    blockpoolI
    D=BP-504949610-10.100.55.95-1493373822794
    layoutVersion=-63
```

In slave nodes (node3 or node4) – you will see blockPoolID for both NNs

```
$ sudo vi /usr/local/hadoop/tmp/dfs/data/current
    BP-489835335-10.100.55.92-1493373813213 VERSION
    BP-504949610-10.100.55.96-1493373822794
```

To validate input data

Now, all services are up and running, including two NNs. To view the available blocks in each NN, go to the following address in browser:

```
node1:50070        // displays list of DNs details that NN1 manage
node5:50070        // displays list of DNs details that NN2 manage
```

When you interact with HDFS via commands, you have to explicitly issue in node1 or node5. If you issue commands in slave nodes, you have to explicitly specify which NN you want to interact with. Example:

In node1,

```
$ hdfs dfs -ls /              // displays only meta-data of NN1 namespace
$ hdfsd dfs -put input.txt /  // uploads input.txt onto NN1 namespace
```

In node5,

```
$ hdfs dfs -ls /              // displays only meta-data of NN2 namespace
$ hdfsd dfs -put input.txt /  // uploads input.txt onto NN2 namespace
```

In nodes other than NN1 and NN2, you have to specify the URI of particular NN you want to interact with. Example:

In node3,

```
$ hdfs dfs -put input.txt hdfs://node1:10001/
$ hdfs dfs -ls hdfs://node1:10001/
    -rw-r--r-- 3 itadmin supergroup 726663168 2017-
    04-28 15:40 hdfs://node1:10001/input.txt
$ hdfs dfs -put file.txt hdfs://node5:10001/
$ hdfs dfs -ls hdfs://node5:10001/
    -rw-r--r-- 3 itadmin supergroup 33 2017-04-28
    15:38 hdfs://node5:10001/file.txt
```

Generic options can be use to specify which NN you interact with. For example,

```
$ hdfs dfs -D fs.defaultFS=node1:10001 -ls /        or
$ hdfs dfs -fs node1:10001 -ls /
```

Once you have uploaded huge data, you can see their blocks in the following locations

```
$ ls /usr/local/hadoop/tmp/dfs/data/current/BP-489835335-
10.100.55.92-1493373813213/current/finalized/subdir0/
subdir0/
```

```
blk_1073741842 blk_1073741844 blk_1073741846
blk_1073741842_1018.meta blk_1073741844_1020.meta
blk_1073741846_1022.meta
blk_1073741843 blk_1073741845 blk_1073741847
blk_1073741843_1019.meta blk_1073741845_1021.meta
blk_1073741847_1023.meta
```

6.20 NAME NODE HIGH AVAILABILITY (NN HA)

The NN failure can happen in several ways. Sometimes, unplanned events such as machine crash, network failure, etc. cause NN failure. Sometimes, planned maintenance events such as software or hardware upgrades in NN requires NN shutdown. During such events NN become inaccessible and this is widely known as SPOF in HDFSv1. If the NN is down, current edit logs and block locations available in the NN memory is lost despite maintaining SNN. Moreover, bringing up fresh NN from SNN takes several minutes as it involves manual intervention. Moreover, the new NN is put on safe mode until

- loaded with FSImage from SNN into memory.
- applied the recent edit logs from SNN.
- received enough BRs from DNs as block locations are maintained only in the memory of NN.

In safe mode, blocks can be read and downloaded to a local file system, but cannot upload any new file into HDFS and launch jobs. Therefore, the NN is inaccessible for 15-30 minutes roughly due to safe mode. This downtime pulls down the production. HA feature addresses the above problems by providing the option of running two or more NNs in the same cluster: active NN and standby/backup/passive NN. Active NN manages the file system operations while standby NN regularly updates meta-data from active NN. When an active NN fails, standby NN becomes active NN immediately without any manual intervention. This is called auto-failover. It is important to note that there is no concept of the SNN. To adapt NN HA, the HDFS requires some architectural changes. NN must use highly available shared storage to share edit logs. So, standby NN can synchronize with active NN and come up with the latest modification. The DNs must send BRs to both active and standby NN because the block locations are stored only in NN memory, and not persistently stored in the disk.

Failover Controller

The transition from standby NN to active NN, ensuring latest FSImage, edit logs and block locations in memory is called failover. Different failover controllers are:

Graceful failover – failover is initiated manually for routine maintenance.

Automatic failover – failover is initiated automatically in case of unplanned events such as the NN failure, network failure. To achieve this, we need a software like a ZooKeeper to achieve auto failover process.

6.20.1 META-DATA SHARING METHODS

There are two ways to achieve highly shared storage to sync meta-data between active and standby NNs: Quorum Journal Manager and NFS.

1. Quorum Journal Manager (QJM)

Both NN communicates through a set of daemons called Journal Nodes (JN) in order to synch meta-state (FSImage). Minimum three (preferably an odd number) JNs should exist in HDFS cluster as shown in Figure 6.7. It is not recommended to launch these services in NN, RM, and slave nodes as it will occupy some resources all the time. However, it is possible to co-locate with Hadoop daemons for practice.

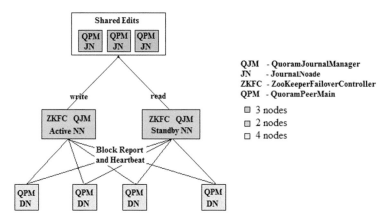

FIGURE 6.7   NN HA architecture.

Every NN runs a process called QJM (background process), which writes logs and meta-data to all JNs. Information written in JNs is consumed periodically by standby NN to keep its file system namespace in sync with active NN. A journal log is considered successfully written only when it is written to the majority of the JNs. Only one of the NNs can achieve this quorum write. JNs follow the ring topology, where JNs are connected in ring form. A JN updates the edit logs and passes to other JNs in the ring. This provides fault-tolerance in case of any JN failure. During failover, the standby NN makes sure that it has updated meta-data from JNs before becoming the new active NN. This makes the current namespace state synchronized with the state before failover. Please note that QJM does not use ZooKeeper. ZooKeeper is mainly used for NN HA to elect new active NN. Once the election is over, standby NN is switched to new active NN in few seconds as standby NN has latest FSImage, edit logs, and up-to-date block mapping readily available in memory.

ZooKeeper (ZK)

It is a software that provides synchronization and coordination among distributed services. The ZK ensures only one NN is active at any point in time. There is no human intervention required switching from standby NN to active NN, unlike HDFSv1. Both active and standby NN must be up-to-date at the moment. This allows a fast failover process. The ZK has the following sub-components, as shown in Figure 6.7 and Figure 6.8.

1. ZooKeeper Failover Controller (ZKFC) – It is responsible for monitoring active NN and initiating automatic failover if any failure. Every NN runs a ZKFC process.
2. QuorumPeerMain – runs in all DNs and JNs to monitor which NN is active at present.

The DNs send BR and HB to both active and standby NN. However, edit log is maintained only in active NN (which writes edit logs in JNs or shared storage). To achieve this, QPM locks the current active NN in all DNs. So, The DNs do not send BR and HB to standby NN. Similarly, QPM locks active NN in JNs. Therefore, only active NN can write its edit logs in JNs. There may be a rare situation where active and standby NN fail together. At this moment, new NN can be introduced manually by recovering FSImage, edit logs, block mapping from highly available shared storage.

NN fencing

Sometimes, during auto-failover, it is impossible to make sure that active NN stopped running before standby NN takes over due to many reasons.

- Slow network or a network partition can trigger failover transition although previous NN is still active.
- If the ZKFC failed in both NN, both NNs would think they are active.

So, once failover happened, DNs might observe two active NNs. This is called split-brain scenario, which may lead to FSImage corruption. So, it is still possible for previous NN to serve stale read requests. In order to avoid the split-brain scenario, before switching from standby NN to active NN, current active NN daemon is killed, and then switching starts. This is called fencing. Another way of fencing is "Shoot The Other Node In The Head" (STONITH), which forcibly powers down the failed NN machine. The failover process is highly transparent to the user. QJM allows only one NN to be the writer at any point in time. Whenever the NN becomes active NN, it generates the epoch number. The NN that has a higher epoch number can be a writer to JNs. This prevents the file system meta-data from getting corrupted. While using NFS, we need a different mechanism to make sure only current active NN writes meta-data.

2. Shared Storage (NFS)

Active NN shares FSImage and edit logs using shared storage like NFS, from which passive NNs can read, as shown in Figure 6.8. However, the problem is, it is not always certain that active NN writes all modifications to NFS. Sometimes, due to a local network problem, some update would have been missed, and standby NN would not be aware of it. Therefore, it is always recommended to use QJM than NFS. During failover, firstly, standby NN updates all meta-data from NFS. Then, it takes the responsibility of active NN.

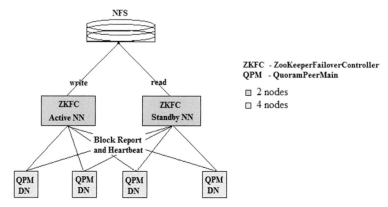

FIGURE 6.8   HA with shared storage.

Hardware requirements for HA deployment

Both active and standby NN should have the equivalent hardware capability.

- The JN daemon is relatively lightweight, so these daemons may reasonably be co-located with NN or RM or DN. Preferably, it is done with DN as master nodes tend to be busier.

- There must be at least three JNs since edit log modifications must be written to a majority of JNs. This will allow the system to tolerate the failure of many JNs. You may also run more than 3 JNs, but in order to increase the number of failures the system can tolerate, you should run an odd number of JNs (3, 5, 7). Note that while running "n" JNs, the system can tolerate at most (n-1)/2 failures and continue to function normally.

6.20.2   NN HA DEPLOYMENT WITH JN

Consider five nodes (PM/VM) as given in Table 6.7 to install different daemons in different nodes. Active NN is set in node1, and standby NN is set in node5. Follow the requirements given in Section 3.1 for all the nodes in the cluster before beginning installation. Refer Figure 3.5 and Figure 5.3 to visualize how Hadoop 2.x multi-node is set up on a cluster of PMs and a cluster of VMs. Refer Tables 3.2 and 3.3 to know the possible cluster configuration for Hadoop. Values such as IP, hostname, and domain name in Table 6.7 are based on my cluster. You have to fill this table according to your cluster environment. Username must be the same in all the nodes in the cluster. I assume username for all the nodes "**itadmin**." Moreover, the hostname must be different in all the nodes. Refer Section 6.2 for some basic requirements to achieve a multi-node cluster. Try multi-node installation before starting NN HA installation.

TABLE 6.7   HDFS NN HA

PM/VM	Daemon	IP	Hostname	Domain name
node1	Active NN + QJM + ZKFC	10.100.55.92	ubuntu1	node1
node2	RM + JN + QPM	10.100.55.93	ubuntu2	node2
node3	Slave1 (DN+NM) + JN + QPM	10.100.55.94	ubuntu3	node3
node4	Slave2 (DN+NM) + JN + QPM	10.100.55.95	ubuntu4	node4
node5	Standby NN + QJM + ZKFC	10.100.55.96	ubuntu5	node5

6.20.2.1   *OS TYPE AND JAVA VERSION IN ALL PHYSICAL/VIRTUAL NODES*

1. OS type must be same in all nodes (either 32-bit OS or 64-bit OS, but not mixed up). Use the following command to check OS type. If it is not the same, it is better to install the right version.

    $ uname -a

    Result for 32-bit Ubuntu will be as below:

    ```
    Linux discworld 2.6.38-8-generic #42-Ubuntu SMP Mon
    Apr 11 03:31:50 UTC 2011 i686 i686 i386 GNU/Linux
    ```

    Result for 64-bit Ubuntu will be as below:

    ```
    Linux ubuntu 4.4.0-104-generic #127-Ubuntu SMP
    Mon Dec 11 12:16:42 UTC 2017 x86_64 x86_64 x86_64
    GNU/Linux
    ```

    Alternatively, you can use the following command

    $ uname -i

    ```
    x86_64
    ```

2. Install Java in all the nodes with the same version (OpenJDK/Oracle JDK) and architecture (32-bit/64-bit). Refer Section 3.1 to install Java. To check java version and architecture,

    $ java -version

    ```
    openjdk version "1.8.0_131"
    OpenJDK Runtime Environment (build
    1.8.0_131-8u131-b11-2ubuntu1.16.04.3-b11)
    OpenJDK 64-Bit Server VM (build 25.131-b11, mixed
    mode)
    ```

    $ javac -version

    ```
    javac 1.8.0_131
    ```

    or go to /usr/lib/jvm

    $ ls /usr/lib/jvm

    Result for 32-bit java will be as below:

    ```
    java-1.8.0-openjdk-i386 java-8-openjdk-i386
    ```

    Result for 64-bit Ubuntu:

    ```
    java-1.8.0-openjdk-amd64 java-8-openjdk-amd64
    ```

Caution:

1. Username must be the same in all the nodes. I have set "**itadmin**" as the username for all the nodes in my cluster. So, create a common username in all nodes in the cluster using the following command.

    ```
    $ sudo adduser itadmin      // leave empty for all the fields
    $ sudo adduser itadmin sudo
    ```

    Now, username itadmin is created. Log into itadmin and do the rest.

2. Another important restriction is, the hostname of the nodes must be unique across the cluster; that is, the hostname should not be identical in the cluster. To change the hostname to ubuntu1 in your machine without restarting, use the following commands

    ```
    $ sudo vi /etc/hostname
        ubuntu1
    $ sudo vi /etc/hosts
        127.0.0.1    localhost
        127.0.1.1    ubuntu1
    $ sudo sysctl kernel.hostname=ubuntu1
    ```

    Do this in all nodes and create unique hostname as given in Table 6.7.

6.20.2.2   *INSTALLATION STEPS*

Step 1: Set the domain name in all five nodes in the cluster

Go to /etc/hosts to set DNS. After setting the domain name, you can use domain names instead of IP address. Comment 127.0.*.1 and enter IP, domain name, and hostname of each node in a line with tab space separator (refer Table 6.7).

```
$ sudo vi /etc/hosts
    #127.0.0.1  localhost
    #127.0.1.1  ubuntu1
    # IP            domain_name        host_name
    10.100.55.92        node1          ubuntu1
    10.100.55.93        node2          ubuntu2
    10.100.55.94        node3          ubuntu3
    10.100.55.95        node4          ubuntu4
    10.100.55.96        node5          ubuntu5
```

Step 2: Set passwordless communication

Create a key in NN1, and NN2 and send to all other nodes in the cluster. Similarly, create a key in RM and send to all other nodes.

Generate key in node1

```
$ ssh-keygen or ssh-keygen -t rsa       // leave empty for name, password
$ ls ~/.ssh/
$ cat ~/.ssh/id_rsa.pub >> ~/.ssh/authorized_keys
$ chmod 0700 ~/.ssh/authorized_keys
```

Send keys from node1 to all other nodes in the cluster: the syntax is

```
$ ssh-copy-id -i ~/.ssh/id_rsa.pub username@IP
$ ssh-copy-id -i ~/.ssh/id_rsa.pub itadmin@node2
$ ssh-copy-id -i ~/.ssh/id_rsa.pub itadmin@node3
$ ssh-copy-id -i ~/.ssh/id_rsa.pub itadmin@node4
$ ssh-copy-id -i ~/.ssh/id_rsa.pub itadmin@node5
```

Similarly, generate key in node2, node5 and send to all other nodes in the cluster.

Step 3 to Step 9 are done in every node in the cluster

Step 3: Download Hadoop 2.7.0 and move into hadoop folder

https://archive.apache.org/dist/hadoop/core/

https://archive.apache.org/dist/hadoop/core/hadoop-2.7.0/hadoop-2.7.0.tar.gz

```
$ tar -zxvf hadoop-2.7.0.tar.gz
$ sudo cp -r hadoop-2.7.0 /usr/local/hadoop
```

Step 4: Download ZK and move into zookeeper folder

https://archive.apache.org/dist/zookeeper/zookeeper-3.4.6/zookeeper-3.4.6.tar.gz

```
$ wget https://archive.apache.org/dist/zookeeper/zookeeper-3.4.6/
zookeeper-3.4.6.tar.gz
$ tar -zxvf zookeeper-3.4.6.tar.gz
$ sudo cp -r zookeeper-3.4.6 /usr/local/zookeeper
```

Step 5: Set path to Hadoop and ZK

```
$ vi .bashrc
```

```
export HADOOP_HOME=/usr/local/hadoop
export PATH=$PATH:$HADOOP_HOME/bin
export PATH=$PATH:$HADOOP_HOME/sbin
export HADOOP_HDFS_HOME=$HADOOP_HOME
export HADOOP_MAPRED_HOME=$HADOOP_HOME
export HADOOP_COMMON_HOME=$HADOOP_HOME
export YARN_HOME=$HADOOP_HOME
export HADOOP_CONF_DIR=$HADOOP_HOME/etc/hadoop
export HADOOP_COMMON_LIB_NATIVE_DIR=$HADOOP_HOME/
lib/native
export HADOOP_OPTS="-Djava.library.
path=$HADOOP_HOME/lib"
export ZOOKEEPER_HOME=/usr/local/zookeeper
export PATH=$PATH:$ZOOKEEPER_HOME/bin
```
$ source.bashrc
$ $PATH

Step 6: Configure Java_Home, disable IPv6 in hadoop-env.sh

$ sudo vi /usr/local/hadoop/etc/hadoop/hadoop-env.sh
```
export JAVA_HOME= /usr/lib/jvm/
java-8-openjdk-amd64
export HADOOP_OPTS=-Djava.net.
preferIPv4Stack=true
```
$ hadoop version

Step 7: Edit configuration files

To set up NN
$ sudo vi /usr/local/hadoop/etc/hadoop/core-site.xml
```
<configuration>
   <property>
      <name>fs.defaultFS</name>
      <value>hdfs://ha-cluster</value>
   </property>
   <property>
      <name>hadoop.tmp.dir</name>
      <value>/usr/local/hadoop/tmp</value>
   </property>
```

```
    <property>
        <name>dfs.journalnode.edits.dir</name>
        <value>/usr/local/hadoop/tmp/jn</value>
    </property>
</configuration>
```

To set up HDFS-related service – to set NNs, JNs, QPMs
$ sudo vi /usr/local/hadoop/etc/hadoop/hdfs-site.xml

```
    <configuration>
        <property>
            <name>dfs.permissions</name>
            <value>false</value>
        </property>
        <property>
            <name>dfs.nameservices</name>
            <value>ha-cluster</value>
        </property>
        <property>
            <name>dfs.ha.namenodes.ha-cluster</name>
            <value>node1,node5</value>
        </property>
    <property>
            <name>dfs.namenode.rpc-address.ha-cluster.
            node1</name>
            <value>node1:10001</value>
        </property>
        <property>
            <name>dfs.namenode.rpc-address.ha-cluster.
            node5</name>
            <value>node5:10001</value>
        </property>
        <property>
            <name>dfs.namenode.http-address.ha-cluster.
            node1</name>
            <value>node1:50070</value>
        </property>
        <property>
```

```
      <name>dfs.namenode.http-address.ha-cluster.
      node5</name>
      <value>node5:50070</value>
   </property>
   <property>
      <name>dfs.namenode.shared.edits.dir</name>
      <value>qjournal://node2:8485;node3:8485;
      node4:8485/ha-cluster</value>
   </property>
   <property>
      <name>dfs.client.failover.proxy.provider.
      ha-cluster</name> <value>org.apache.hadoop.
      hdfs.server.namenode.ha.ConfiguredFailoverProx
      yProvider</value>
   </property>
   <property>
      <name>dfs.ha.automatic-failover.enabled</name>
      <value>true</value>
   </property>
   <property>
      <name>ha.zookeeper.quorum</name>
      <value>node2:2181,node3:2181,node4:2181</
      value>
   </property>
   <property>
      <name>dfs.ha.fencing.methods</name>
      <value>sshfence</value>
   </property>
   <property>
      <name>dfs.ha.fencing.ssh.private-key-files</
      name>
      <value>/home/itadmin/.ssh/id_rsa</value>
   </property>
</configuration>
```

`dfs.permissions` – allows any users to read/write in HDFS.

`ha.zookeeper.quorum` – includes nodes that run QPM.

`dfs.namenode.shared.edits.dir` – includes nodes that run only JNs.

To set up ZK

$ sudo cp -r /usr/local/zookeeper/conf/zoo_sample.cfg /usr/local/
zookeeper/conf/zoo.cfg
$ sudo mkdir /usr/local/zookeeper/tmp
$ sudo vi /usr/local/zookeeper/conf/zoo.cfg // to indidcate node that will
run QPMs

```
tickTime=2000
initLimit=10
syncLimit=5
dataDir=/usr/local/zookeeper/tmp
clientPort=2181
#following includes node that runs QPM
server.1=node2:2888:3888
server.2=node3:2888:3888
server.3=node4:2888:3888
```

To set up YARN

$ sudo vi /usr/local/hadoop/etc/hadoop/yarn-site.xml

```
<configuration>
  <property>
     <name>yarn.nodemanager.aux-services</name>
     <value>mapreduce_shuffle</value>
  </property>
  <property>
     <name>yarn.nodemanager.aux-services.mapreduce.
     shuffle.class</name> <value>org.apache.hadoop.
     mapred.ShuffleHandler</value>
  </property>
  <property>
     <name>yarn.resourcemanager.hostname</name>
     <value>ubuntu2</value>
  </property>
  <property>
     <name>yarn.resourcemanager.address</name>
     <value>node2:8030</value>
  </property>
  <property>
```

```
        <name>yarn.resourcemanager.resource-tracker.
        address</name>
        <value>node2:8031</value>
    </property>
    <property>
        <name>yarn.resourcemanager.scheduler.address</
        name>
        <value>node2:8032</value>
    </property>
    <property>
        <name>yarn.resourcemanager.admin.address</
        name>
        <value>node2:8033</value>
    </property>
    <property>
        <name>yarn.resourcemanager.webapp.address</
        name>
        <value>node2:8088</value>
    </property>
    <property>
        <name>yarn.nodemanager.disk-health-checker.
        min-healthy-disks</name>
            <value>0.0</value>
    </property>
    <property>
        <name>yarn.nodemanager.disk-health-checker.
        max-disk-utilization-per-disk-percentage</
        name>
        <value>98.5</value>
    </property>
</configuration>
```

To set up MR

$ sudo cp /usr/local/hadoop/etc/hadoop/mapred-site.xml.template /usr/
local/hadoop/etc/hadoop/mapred-site.xml
$ sudo vi /usr/local/hadoop/etc/hadoop/mapred-site.xml

```
    <configuration>
```

```
<property>
   <name>mapreduce.framework.name</name>
   <value>yarn</value>
</property>
<property>
   <name>mapreduce.jobhistory.address</name>
   <value>node2:10020</value>
</property>
<property>
   <name>mapreduce.jobhistory.webapp.address</
   name>
   <value>node2:19888</value>
</property>
<property>
   <name>mapreduce.jobhistory.
   intermediate-done-dir</name>
   <value>/usr/local/hadoop/tmp</value>
</property>
<property>
   <name>mapreduce.jobhistory.done-dir</name>
   <value>/usr/local/hadoop/tmp</value>
</property>
</configuration>
```

To set up slave (DN and NM)

$ sudo vi /usr/local/hadoop/etc/hadoop/slaves

```
node3
node4
```

Step 9: Grant ownership and access rights

$ sudo chown -R itadmin /usr/local/
$ sudo chmod -R 777 /usr/local/

Step 10: Start JNs in node2, node3, node4

$ hadoop-daemon.sh start journalnode
$ jps   // you will see JN running and can see directories
$ ls /usr/local/hadoop/tmp/

Step 11: Format and start active NN in node1 and standby NN in node5

Run the following commands in node1.

```
$ hdfs namenode -format
$ hadoop-daemon.sh start namenode
$ jps                // Active NN is running
```

You need not format standby NN in node5. Just run the following commands in node5 to copy meta-data from active NN to standby NN.

```
$ hdfs namenode -bootstrapStandby  // make sure no exception arrives
$ hadoop-daemon.sh start namenode
$ jps                          // Standby NN is running
```

After starting NN, you can check /usr/local/hadoop/tmp/jn directory in node2, node3, and node4 to see the details of active NN and HDFS cluster.

```
$ vi /usr/local/hadoop/tmp/jn/ha-cluster/current/VERSION
    #Sat Apr 29 16:35:33 IST 2017
    namespaceID=241088241
    clusterID=CID-aaa54bcd-571a-42ee-b005-
    a127f3cf4f8a
    cTime=0
    storageType=JOURNAL_NODE
    layoutVersion=-63
```

Step 12: Set up ZK id and start ZK

You have to follow the order as given in /usr/local/zookeeper/conf/zoo.cfg

In node2,

```
$ sudo vi /usr/local/zookeeper/tmp/myid
    1
$ sudo chown -R itadmin /usr/local/zookeeper
$ sudo chmod -R 777 /usr/local/zookeeper
$ zkServer.sh start
$ jps                // you will see QuorumPeerMain running
```

In node3,

```
$ sudo vi /usr/local/zookeeper/tmp/myid
    2
$ sudo chown -R itadmin /usr/local/zookeeper
$ sudo chmod -R 777 /usr/local/zookeeper
```

```
$ zkServer.sh start
$ jps                          // you will see QuorumPeerMain running
```

In node4,

```
$ sudo vi /usr/local/zookeeper/tmp/myid
       3
$ sudo chown -R itadmin /usr/local/zookeeper
$ sudo chmod -R 777 /usr/local/zookeeper
$ zkServer.sh start
$ jps                          // you will see QuorumPeerMain running
```

After QPM started in node3, node4, and node5, you can see some files in the following location.

```
$ ls /usr/local/zookeeper/tmp/
       myid version-2 zookeeper_server.pid
$ zkServer.sh status           // to check QPM is leader or follower
```

You can see who is leader and follower among ZK services running in JNs.

Step 13: Start DN+NM in node3 and node4

```
$ hadoop-daemon.sh start datanode
$ yarn-daemon.sh start nodemanager
$ jps
```

Step 14: Formatting ZK and starting ZKFC

Run the following commands in node1 to format ZK and start ZKFC process.

```
$ hdfs zkfc -formatZK
$ hadoop-daemon.sh start zkfc
$ jps
```

Run the following in node5 to start the ZKFC process (need not format ZK again).

```
$ hadoop-daemon.sh start zkfc
$ jps
```

Now, check the status (active or standby) of each NN using the following command in any node of the Hadoop cluster.

```
$ hdfs haadmin -getServiceState node1
       17/05/05 17:53:18 WARN util.NativeCodeLoader:
       Unable to load native-hadoop library for your
```

```
platform... using builtin-java classes where
applicable
active
```

$ hdfs haadmin -getServiceState node5

```
17/05/05 17:54:48 WARN util.NativeCodeLoader:
Unable to load native-hadoop library for your
platform... using builtin-java classes where
applicable
Standby
```

You can check the status via WUI. It will show whether the NN is active or on standby.

node1:50070 (Figure 6.9)

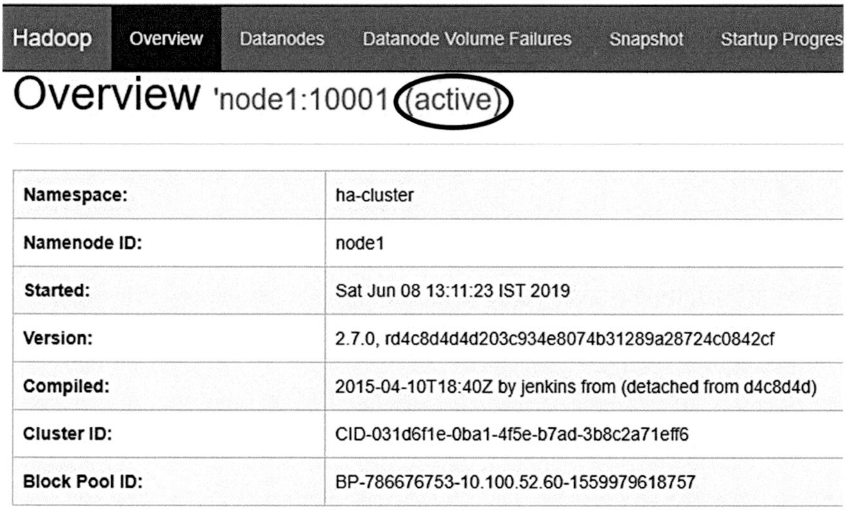

FIGURE 6.9   WUI for active NN.

node5:50070 (Figure 6.10)

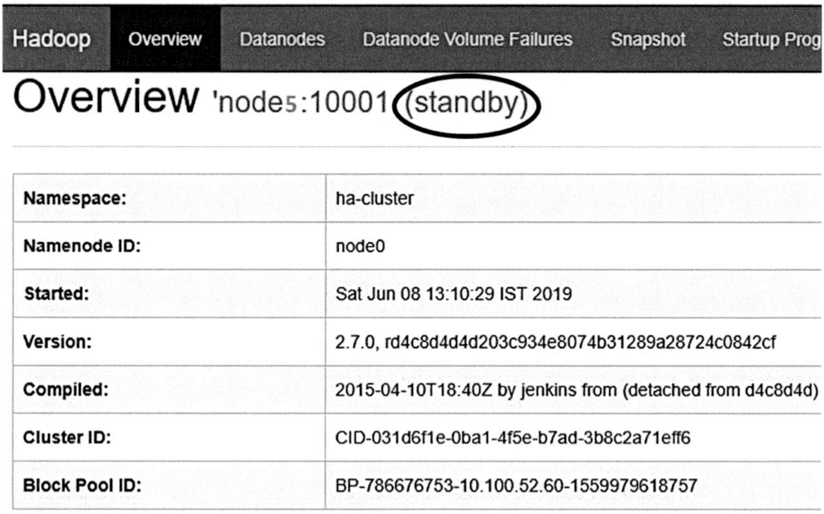

| Hadoop | Overview | Datanodes | Datanode Volume Failures | Snapshot | Startup Prog |

Overview 'node5:10001 (standby)'

Namespace:	ha-cluster
Namenode ID:	node0
Started:	Sat Jun 08 13:10:29 IST 2019
Version:	2.7.0, rd4c8d4d4d203c934e8074b31289a28724c0842cf
Compiled:	2015-04-10T18:40Z by jenkins from (detached from d4c8d4d)
Cluster ID:	CID-031d6f1e-0ba1-4f5e-b7ad-3b8c2a71eff6
Block Pool ID:	BP-786676753-10.100.52.60-1559979618757

FIGURE 6.10   WUI for standby NN.

Step 15: Launch RM in node2,

$ yarn-daemon.sh start resourcemanager
$ jps

Step 16: Failover process

To test how failover takes place and standby NN becomes active NN, let us kill the active NN daemon in node1. So, node5 should become active NN as NN in node1 is down.

$ jps   // to get process ID of NN service to kill
$ sudo kill -9 NN_daemon_ID

Now, check the status of NNs.

$ hdfs haadmin -getServiceState node1   // throws exception as NN is not running
$ hdfs haadmin -getServiceState node5

```
17/05/05 17:54:48 WARN util.NativeCodeLoader:
Unable to load native-hadoop library for your
platform... using builtin-java classes where
applicable
Active
```

To perform a graceful (manual) failover process, execute the following command

```
$ hdfs haadmin -failover -forceactive activeNN standbyNN
$ hdfs haadmin -failover -forceactive node1 node5
```

You can check the status via WUI. It will show whether the NN is active or standby.

node1:50070 (Figure 6.11)

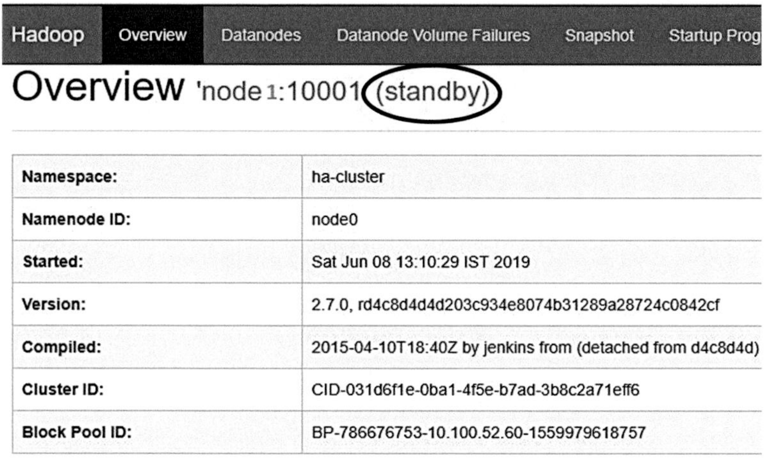

FIGURE 6.11 WUI for active NN.

6.21 HDFS PERSISTENT DATA STRUCTURES

A Hadoop administrator should have some basic understanding of how HDFS components (NN, SNN, and DN) organize their persistent data in disk. The HDFS daemons maintain directories and files in a tree structure. A default location can be set using hadoop.tmp.dir property in core-site.xml for NN, SNN, and DN to store its data persistently. We usually provide /usr/local/hadoop/tmp location in our examples. After formatting HDFS and starting HDFS services in pseudo-distributed mode, you will get the following directories.

```
$ tree -d /usr/local/hadoop/tmp/dfs
    ├── data
    ├── name
    └── namesecondary
```

/tmp/dfs/name directory is created when you format HDFS. /tmp/dfs/ data directory is created when DN daemon is launched. /tmp/dfs/names-econdary is created when SNN is launched. Let us discuss these directories and its structure one by one.

6.21.1   */TMP/DFS/NAME*

The NN directory location can be specified explicitly using dfs.namenode. name.dir property in hdfs-site.xml. After NN daemon started, you will find a file and directory under name directory.

$ ls /usr/local/hadoop/tmp/dfs/name/

```
        current in_use.lock
```

$ tree /usr/local/hadoop/tmp/dfs/name

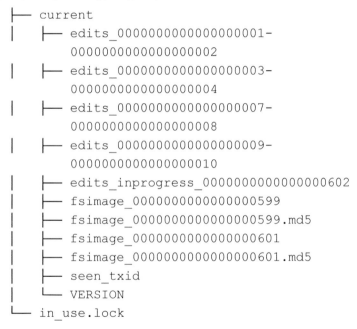

```
        ├── current
        │   ├── edits_0000000000000000001-
        │       0000000000000000002
        │   ├── edits_0000000000000000003-
        │       0000000000000000004
        │   ├── edits_0000000000000000007-
        │       0000000000000000008
        │   ├── edits_0000000000000000009-
        │       0000000000000000010
        │   ├── edits_inprogress_0000000000000000602
        │   ├── fsimage_0000000000000000599
        │   ├── fsimage_0000000000000000599.md5
        │   ├── fsimage_0000000000000000601
        │   ├── fsimage_0000000000000000601.md5
        │   ├── seen_txid
        │   └── VERSION
        └── in_use.lock
```

VERSION – contains version of HDFS that is running.

$ cat /usr/local/hadoop/tmp/dfs/name/current/VERSION

```
        #Mon Jun 05 15:23:49 IST 2017
        namespaceID=420371433
        clusterID=CID-9a7b7118-5c29-4508-8f4e-
        1cfd3a2970f7
```

```
cTime=0
storageType=NAME_NODE
blockpoolI
D=BP-519560340-10.100.52.145-1496656429357
layoutVersion=-63
```

namespaceID – unique identifier for file system namespace is created when HDFS is formatted. In HDFS federation, there are multiple NNs managing its own namespace.

clusterID – unique identifier that specifies the HDFS cluster.

cTime – time at which HDFS was created.

storageType – denotes HDFS daemon that manages storage in respective node. In slave node, it is DN.

blockpoolID – unique identifier for block pool containing blocks managed by a NN.

layoutVersion – negative integer to represent the version of HDFS meta-data format. This version number is no way related to Hadoop distribution. Whenever the layout changes, version number is decremented. It means that HDFS needs to be upgraded, else DNs will not be linked with the NN.

in_use.lock – NN uses this file to lock NN storage directory from rewriting by other NN at the same time that leads to data corruption and loss.

Edit logs and FSImage

A client performing read/write operation is called a transaction in HDFS. A unique transaction ID is assigned for every file system modification. Every transaction after latest checkpointing is recorded in the edit logs before applying onto the file system. Whenever a transaction is done, it is written as edit file with a name containing transactionID in edit logs. Complete HDFS state at any point of time is called FSImage, which contains all the edit files in edit logs that are finalized. The internal representation of a file/directory meta-data is called inode, which records meta-data such as file name, number of replications, last modified, access permissions, number of blocks, and its locations. For directories, the modification time, permissions, and quota meta-data are stored.

An FSImage file does not record which blocks are stored in which DN. The NN gets a block location information using periodic BR from DN and maintains in memory itself. After a certain size of edit logs in memory, it should be merged with old FSImage file stored in the disk to be up to date. This is called checkpointing. This process cannot take place in NN memory as NN is busy serving DNs and clients. Therefore, we need one more node, called SNN in HDFSv1. Only one edit file will be open for recording transactions, and its name is prefixed with edits_inprogress. Example:edits_ inprogress_0010. While passive NN becomes active NN, passive NN can only read up to the finalized edit segments. It will not be up to date with the current "edit in progress" file. So, when failover happens, the current "edit in progress" file should be merged with the latest FSImage version before passive NN becomes active NN.

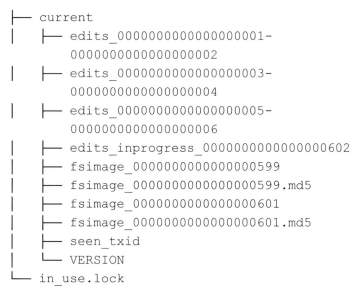

```
├── current
│    ├── edits_0000000000000000001-
│         0000000000000000002
│    ├── edits_0000000000000000003-
│         0000000000000000004
│    ├── edits_0000000000000000005-
│         0000000000000000006
│    ├── edits_inprogress_0000000000000000602
│    ├── fsimage_0000000000000000599
│    ├── fsimage_0000000000000000599.md5
│    ├── fsimage_0000000000000000601
│    ├── fsimage_0000000000000000601.md5
│    ├── seen_txid
│    └── VERSION
└── in_use.lock
```

6.21.2   /TMP/DFS/NAMESECONDARY

The NN suffers from a SPOF. If the NN is down, the entire HDFS becomes inaccessible. So, we introduce SNN to backup FSImage file. If the NN crashed, the administrator can bring a new node and restore FSImage from the SNN. The added advantage of SNN is to perform the checkpointing, which is more compute-intensive to do at NN.

-importCheckpoint option will load NN meta-data from the latest checkpoint. When the SNN starts running, it manages a copy of FSImage in memory from the NN. When the number of edit files reaches one million transactions (dfs. namenode. checkpoint.txns) or every one hour (dfs.namenode.checkpoint. period in seconds), edit logs (all edit files) are transferred over HTTP to SNN.

The SNN merges the current edit logs with the previous FSImage to create new FSImage. This merging process is called checkpointing. Checkpointed file (new FSImage file) contains .ckpt extension. Once new FSImage is created, the SNN sends to NN, which replaces old FSImage with new FSImage and flushes out old edit logs. The number of transactions is checked every minute (dfs.namenode. checkpoint.check.period in seconds) to trigger checkpointing either by using configuration properties or manually by HDFS administration commands.

Until checkpointing process is over, HDFS is on safe mode which allows only read operation. The NN also remembers the last transactionID of the last checkpoint process using seen_txid file. Once the NN received the new FSImage and come out of safe mode, edit files are created again from last seen_txid. It is recommended that the SNN hardware configuration should match the NN configuration to have seamless performance. The SNN meta-data contains one directory and one file as in the NN. Its location can be specified using dfs.namenode.checkpoint.dir property in hdfs-site.xml file. We see the location we have set in our examples for storing meta-data.

$ ls /usr/local/hadoop/tmp/dfs/namesecondary/

```
current  in_use.lock
```

$ tree /usr/local/hadoop/tmp/dfs/namesecondary

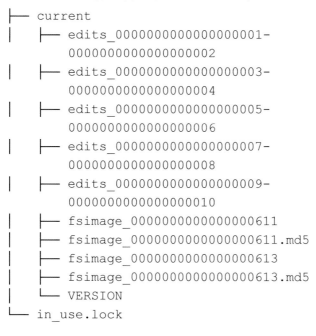

```
├── current
│   ├── edits_0000000000000000001-
│   │   0000000000000000002
│   ├── edits_0000000000000000003-
│   │   0000000000000000004
│   ├── edits_0000000000000000005-
│   │   0000000000000000006
│   ├── edits_0000000000000000007-
│   │   0000000000000000008
│   ├── edits_0000000000000000009-
│   │   0000000000000000010
│   ├── fsimage_0000000000000000611
│   ├── fsimage_0000000000000000611.md5
│   ├── fsimage_0000000000000000613
│   ├── fsimage_0000000000000000613.md5
│   └── VERSION
└── in_use.lock
```

`in_use.lock file` contains the last transactionID before checkpointing process.

$ cat /usr/local/hadoop/tmp/dfs/namesecondary/in_use.lock
```
    1023
```

6.21.3   /TMP/DFS/DATA

Unlike NN, DNs create their storage directories automatically upon DN daemon startup. The DN directory location can be specified using dfs. datanode.data.dir property in `hdfs-site.xml` file. We shall see the location set in our examples for storing data blocks and other meta-data.

$ ls /usr/local/hadoop/tmp/dfs/data/
```
    current in_use.lock
```
$ tree /usr/local/hadoop/tmp/dfs/data/
```
        ├── current
        │   ├── BP-519560340-10.100.52.145-1496656429357
        │   │   ├── current
        │   │   │   ├── finalized
        │   │   │   │   └── subdir0s
        │   │   │   │   └── subdir0
        │   │   │   │   ├── blk_1073741825
        │   │   │   │   ├── blk_1073741825_1001.meta
        │   │   │   │   ├── blk_1073741826
        │   │   │   │   ├── blk_1073741826_1002.meta
        │   │   │   │   ├── blk_1073741827
        │   │   │   │   ├── blk_1073741827_1003.meta
        │   │   │   │   ├── blk_1073741828
        │   │   │   ├── rbw
        │   │   │   ├── lazyPersist
        │   │   │   └── VERSION
        │   │   └── tmp
        │   │   └── dncp_block_verification.log.curr
        │   │   └── dncp_block_verification.log.prev
        │   └── VERSION
        └── in_use.lock
```

The HDFS blocks are stored as files in the local file system with a prefix "blk." This file contains raw data blocks of an original file. Each block file is associated with a meta-data file with ".meta" extension. This file comprises header and checksum for error detection. Each block belongs to a block pool, and each block pool has its storage directory. When the number of blocks in a directory goes beyond a limit (dfs.datanode.numblocks, by default 64), DN creates a new subdirectory, in which new blocks and their accompanying meta-data are stored. The aim is to maintain high fan-out, which means that every DN has to have almost the same number of sub-directories, such that the directory tree is maintained at the same level.

By taking this measure, DN ensures that there is a manageable number of block files per directory, which avoids the problems that most OS encounter when there are a large number of files (hundreds of thousands) in a single directory. Blocks are stored in a round-robin fashion if we specify multiple directory paths for dfs.data node.data.dir property.

BP-RandomInteger-NN_IP-creationTime – The naming convention of this directory is significant and constitutes a form of cluster meta-data. "BP" stands for "block pool," and denotes a set of blocks belonging to a single namespace. In the case of a federated deployment, there will be multiple "BP" sub-directories, one for each NN. Other components followed by BP are a random integer, IP address of the NN that created the block pool, and time created.

finalized/rbw – rbw stands for replica being written. Both finalized and rbw contains a directory structure for block storage. This holds numerous block files and its.meta-data.

dncp_block_verification.log – This file tracks the time at which data is verified with its checksum. The last verification time is significant in deciding how to prioritize the subsequent verification work. The DNs order its background block verification work in ascending order of last verification time. This file is rolled periodically, so it is typical to see a .curr file (current) and a .prev file (previous).

in_use.lock – This is a lock file held by DN, used to prevent multiple NN from trying to access at the same time.

6.21.4   *JOURNAL NODE (JN)*

In HA deployment, edits are logged to a separate set of daemons called JNs, whose meta-data directory is configured by setting dfs.journalnode.edits.dir.

The JN contains VERSION file, multiple edit files, and an edit_inprogress, just like the NN. However, JN will not have fsimage files or seen_txid. In addition, it contains several other files relevant to HA implementation. These files help prevent a split-brain scenario, in which multiple NN could think they are active and try to write edits.

committed-txid – Tracks last transactionID committed by a NN.

last-promised-epoch – This file contains the "epoch" number, which is a monotonically increasing integer. When a new NN starts as active, it increments the epoch and presents it to JN. The NN which has a larger epoch number acquires a lock to write in JNs.

last-writer-epoch – It is similar to last-promised-epoch, but last-writer-epoch contains the epoch number associated with the writer (NN) which wrote a transaction in the recent past.

paxos – It is a directory containing temporary files used in the implementation of the paxos distributed consensus protocol. This directory will often appear as empty.

6.21.5   SAFE MODE

During the checkpointing process, the NN will be running in safe mode. Safe mode allows HDFS to serve only to read operations and disable write operations for clients. It is because the block location information is not persisted, but maintained in the NN memory. When the NN is restarted or boots up freshly, it waits for all DNs to send BR to build the block location information in the NN memory. Until it receives enough BR from all DNs, the NN will be on safe mode. In safe mode, NN cannot instruct DNs to remove blocks or replicate. The NN gets into safe mode due to the following actions:

- When HDFS is freshly formatted, the NN gets into safe mode until it receives a sufficient number of BRs from DNs.
- During Hadoop upgrade.

Safe mode properties

dfs.namenode.replication.min – the minimum number of replications to be made for a write to be successful. Default is 1.

dfs.namenode.safemode.threshold-pct – percentage of blocks to meet minimum replication before the NN exits safe mode. Default is 0.999. Value

0 or less than 0 stops NN entering into safe mode. Value 1 never allows NN to exit safe mode.

dfs.namenode.safemode.extension – this extends safe mode after the minimum replication condition defined by dfs.namenode.safemode. threshold-pct has been satisfied. The default value is 30000.

Safe mode commands

To check whether the NN is in safe mode, visit WUI of HDFS. Alternatively, you can use the command to verify.

$ hdfs dfsadmin -safemode get

Sometimes, to wait for the NN to exit safe mode before interacting with HDFS.

$ hdfs dfsadmin -safemode wait

An administrator has the ability to make NN enter or leave safe mode at any time. It is sometimes necessary to carry out maintenance or after upgrading the cluster.

$ hdfs dfsadmin -safemode enter/leave

6.21.6   HDFS ADMINISTRATIVE COMMANDS

It is highly crucial for Hadoop cluster administrators to maintain HDFS cluster up and running. Therefore, it becomes essential to test cluster performance and monitor the health of data blocks. Let us see some important administrative activities and commands used.

1. dfsadmin

To find more information about HDFS and blocks, the dfsadmin tool is used. It helps to perform some administrative operations as well.

$ hdfs dfsadmin -command

The commands are

help	- displays a hint on commands used with dfsadmin.
report	- shows filesystem statistics and information about DN.
metasave	- dumps information about blocks into the log file.
safemode	- to enter/leave read-only mode.
saveNamespace	- saves current FSImage and resets edit files, performed in safe mode.
fetchImage	- retrieves the recent FSImage from NN and saves in the local file.

refreshNodes   - updates set of DNs to join the cluster.

2. Filesystem check (fsck)

It is a utility tool to check the health of blocks in HDFS. This tool looks for under or over-replicated blocks.

```
$ hdfs fsck /
    Status: HEALTHY
    Total size:                139135 B
    Total dirs:                22
    Total files:               12
    Total symlinks:            0
    Total blocks (validated):  11 (avg.blocksize12648B)
    Minimally replicated blocks: 11 (100.0 %)
    Over-replicated blocks:    0 (0.0 %)
    Under-replicated blocks:   11 (100.0 %)
    Mis-replicated blocks:     0 (0.0 %)
    Default replication factor: 3
    Average block replication: 1.0
    Corrupt blocks:            0
    Missing replicas:          22 (66.666664 %)
    Number of data-nodes:      1
    Number of racks:           1
    FSCK ended at Thu Jun 15 01:18:19 IST 2017 in 6
    milliseconds
```

fsck recursively explores through the file system namespace from the given path. It deals with NN and retrieves meta-data of blocks and looks for problems or inconsistencies. fsck checks for some conditions as given below:

Over-replicated blocks – Number of replicated blocks is higher than the RF. It is not a problem. However, over-replication wastes the storage space, so it is deleted after some time.

Under-replicated blocks – When the number of replicas of a block does not meet the RF, under-replication happens. It is a serious issue as it determines the amount of parallelism that we can extract. So, under-replication is given higher priority to meet the RF.

Mis-replicated blocks – When block replicas are placed without satisfying rack awareness, mis-replication happens. It is not a serious issue until any node/rack failure happens. The HDFS will automatically re-replicate such blocks to satisfy the rack awareness.

Corrupted blocks – A block is corrupt when all its replicas are corrupt due to the wrong checksum. The checksum is used to find error detection and nothing to do with error correction. Therefore, if a block is corrupt, it is almost lost.

Missing blocks – These are the blocks with no replica anywhere in the cluster.

fsck tool provides an easy way to find out in which DN blocks have been stored. Example:

```
$ hdfs fsck /filename -files -blocks -racks      or
$ hdfs fsck /filename -files -blocks -locations
```

```
/output/part-r-00000 51 bytes, 1 block(s):
Under replicated BP-519560340-10.100.52.145-
1496656429357:blk_1073741840_1016. Target Replicas
is 3 but found 1 replica(s).
0.BP-519560340-10.100.52.145-
1496656429357:blk_1073741840_1016 len=51 repl=1 [/
default-rack/10.100.52.145:50010]
    Status: HEALTHY
    Total size:              51 B
    Total dirs:              1
    Total files:             2
    Total blocks (validated):  1 (avg. block size 51 B)
    Minimally replicated blocks: 1 (100.0 %)
    Over-replicated blocks:    0 (0.0 %)
    Under-replicated blocks:   1 (100.0 %)
    Mis-replicated blocks:     0 (0.0 %)
    Default replication factor: 3
    Average block replication: 1.0
    Corrupt blocks:            0
    Missing replicas:          2 (66.666664 %)
    Number of data-nodes:      1
    Number of racks:           1
```

```
FSCK ended at Thu Jun 15 01:33:44 IST 2017 in 1
milliseconds
```

hdfs -fsck without any arguments displays the entire HDFS. There are some options for this command.

- files → displays the name of the files line by line along with its size, number of blocks and its health.
- blocks → displays the block information of each file.
- racks → shows the DN addresses, rack number for each block.

To delete unwanted blocks
  $ hadoop fsck -delete

3. Meta-data backups

Despite we maintain the SNN/NN HA to overcome SPOF, it would be better to back up a copy of FSImage periodically (hour/day/week). Because losing meta-data causes permanent loss or damage of the file system. To backup most recent FSImage file

  $ hdfs dfsadmin -fetchImage fsimage.backup

4. Data backup

The HDFS achieves fault-tolerance using replication of blocks at the soft-ware-level. However, losing data is still possible when multiple nodes fail at the same time. Indeed, we cannot backup huge data. However, essential data of business/banking that cannot be regenerated is a matter of concern. We use a tool "distcp" for backup from one HDFS cluster to another HDFS cluster or other file systems such as S3.

  $ yarn distcp hdfs://NN1_IP/source_file hdfs://NN2_IP/desti_dir

Users and administrators can take a snapshot of the file system. A snapshot is a read-only copy of a file system at a given point in time. It is very efficient as it does not copy the physical blocks but takes the meta-data that is sufficient to reconstruct the file system. It is not a replace-ment for data backups, but it is useful for data recovery of files that were mistakenly deleted by users. You can customize to backup hourly/ monthly.

6.22   UPGRADING HADOOP VERSION

In VERSION file, if layout version is changed, meta-data is automatically migrated to a format that is compatible with the new version using the upgrade tool. Before upgrading Hadoop cluster with the latest version, shut down all daemons, update the configuration file, and start the new daemons. This process is reversible, so rolling back an upgrade is also straightforward. After every successful upgrade, you should perform a couple of final clean-up steps:

1. Remove the old installation and configuration files from the cluster.
2. Fix any deprecation warnings in your code and configuration.

Hadoop cluster management tools like Cloudera manager and Ambari simplify the upgrade process and also make it easier to do rolling upgrades, where nodes are upgraded in batches, so clients do not experience any service interruptions. Upgrading steps are:

1. Ensure any previous upgrade is finalized before proceeding with the new upgrade.
2. Shutdown YARN and MR daemons.
3. Shutdown HDFS, and back up NN directories.
4. Install the new version of Hadoop in the cluster.
5. Start HDFS with the -upgrade option.
6. Wait until the upgrade is complete.
7. Perform sanity checks on HDFS.
8. Start HDFS, YARN, and MR daemons.
9. Roll-back or finalize the upgrade (optional).

Start the upgrade

To upgrade, run the following command

$ start-all.sh -upgrade

This causes the NN to upgrade its meta-data after placing the previous version in a new directory called dfs.namenode.name.dir. Similarly, DNs upgrade their storage directories, preserving the old copy in a directory. Wait until the upgrade is complete. The upgrade process is not instantaneous, but you can check the progress of an upgrade using dfsadmin.

$ hdfs dfsadmin -upgradeProgress status

```
Upgrade for version -18 has been completed.
The upgrade is not finalized.
```

Check the upgrade

At this stage, you should run some sanity checks on the filesystem. You might choose to put HDFS into safe mode while you are running sanity checks to prevent others from making changes.

```
$ hadoop fsck / -files -locations -blocks
```

Roll-back the upgrade (optional)

If you find that the new version is not working correctly, you may choose to roll-back to the previous version. This is possible only if you have not finalized the upgrade. A roll-back reverts the file system state before the upgrade was performed, so any changes made in the meantime will be lost. In other words, it rolls back to the previous state of the file system.

First, shut down the new daemons:

```
$ stop-dfs.sh
```

Then, start the old version of HDFS with the -rollback option:

```
$ start-dfs.sh -rollback
```

This command gets the NN and DNs to replace their current storage directories with their previous copies.

Finalize the upgrade (optional)

When you are happy with the new version of HDFS, you can finalize the upgrade to remove the previous storage directories. After an upgrade has been finalized, there is no way to roll back to the previous version. The following steps are done before performing another upgrade.

```
$ hdfs dfsadmin -finalizeUpgrade
$ hdfs dfsadmin -upgradeProgress status
    There are no upgrades in progress.
```

The HDFS is now fully upgraded to the new version.

6.23   RESEARCH TOPICS IN HADOOP

Hadoop is one of the cost-effective opensource frameworks for big data processing. Although some advanced tools are coming up to solve different data problems, Hadoop remains active due to its low-cost support nature. So, research in Hadoop is prevalent towards optimizing resource utilization and improving job performance. Moreover, Hadoop is also available as a service from Cloud Service Providers (CSP) such as Microsoft, Amazon, Google, etc. Some of the interesting topics in Hadoop are (but not limited to):

- block and replication placement
- replication policy and protocols
- load balancing policies
- scaling HDFS namespace
- improving YARN and MRAppMaster scheduling performance
- enhancements in speculative execution
- adding the capability to work with encrypted data
- SLA based resource allocation and scheduling
- migration of task and applications
- caching schemes for multi-stage applications
- in-memory sorting and shuffling
- discovering network topology
- identifying and diagnosing hardware that is not functioning correctly
- debugging and performance optimization
- distributed profiler for measuring distributed applications
- integration of virtualization (such as Xen) with Hadoop tools
- exploiting the heterogeneous performance of virtual nodes
- profiling heterogeneous workloads

6.24   HADOOP IN CLOUD

Establishing large scale Hadoop cluster on-premise is not affordable for small scale businesses, educational institutions. Therefore, CSP offers Hadoop and relevant applications as a service on-demand for pay-per-use basis. Example: Microsoft Azure HDInsight, Amazon EMR, Rackspace Hadoop, etc. CSP delivers Hadoop to end users via Infrastructure as a Service (IaaS) on two different platforms: a cluster of dedicated servers and a cluster of virtual servers. Hadoop MR as a service is offered in different flavors.

1. Private Hadoop MR (pay per VM/PM hired)
   – Purchase VMs/PMs from CSP and set up MR manually
   – Hire MR itself as a service

2. Sharing MR service with more than one users (pay per job basis).

The advantages of using Hadoop from the cloud are:

- Need not purchase advanced hardware/licensed software. So, there is no upfront capital investment.
- The CSP itself manages resources.
- Increase or decrease the number of instances (VMs) instantly and dynamically.
- Pay for what you use and only when you use.
- Visualization and data management tools are provided along with Hadoop service for free of cost.

New York Times converted 4 TB of image articles into PDFs with S3 and EC2 instance in AWS cloud using Hadoop and did the work in less than 24 Hours for less than $240. The only drawback is, uploading your big data onto the cloud will take much time and depends on our local network speed. Let us discuss how to access Hadoop from the cloud with Microsoft Azure and Amazon.

6.24.1   HADOOP ON MICROSOFT AZURE

As we mentioned before, there are two ways of obtaining a Hadoop cluster from the cloud.

Way 1: user hires a set of VMs/PMs and set up Hadoop cluster on it

Go to https://azure.microsoft.com/en-in/account/ to create a cloud account in Azure (need an outlook mail and credit card). Then, follow the procedure:

1. Create a network in Azure
2. Create a set of Ubuntu VMs: open required ports such as 50070 and 50030 because, within the cloud cluster all ports are visible, but to access from the public we need to open specific port numbers.
3. Add VMs to the created network.
4. VMs in the cloud typically has two IP: global (public or floating IP), local (private IP). Local IP is used within cloud network, global IP is used to access VMs by users over the Internet.
5. Connect VMs through putty using global IP.
6. Now, follow the same installation procedure we discussed for single node and multi-node implementation.

Way 2: Microsoft Azure HDInsight cloud service

Microsoft offers Hadoop virtual cluster itself as a service (HDInsight). We need to specify how many nodes required in the Hadoop cluster while starting this service. CSP itself does all other Hadoop daemon set up and configurations. As a result, you will have the Hadoop cluster, on which you can launch jobs. Steps are

- Create a storage account in Azure: New → Data services → Storage account → Quick create → URL and location.
- Create an HDInsight account: New → Data services → HDInsight → Quick create → URL, cluster size, password, choose storage account we created now.

HDInsight Advantages

HDInsight provides Ambari for Hadoop cluster management, monitoring with a lot of other additional features. It provides full integration with Microsoft BI stack, SQL Server (RDBMS, SSAS-SQL Server Analysis Server for multi-dimensional data analysis, SSRS-SQL server reporting services), Sharepoint especially for Powerview and Office suite like Excel, Powerpoint, etc. It provides Hive ODBC driver for Excel, PowerPivot for visualization. Data market place is also available in Azure. Example: Zip code information is available as a service.

Case study

343 Studios produced a HALO 4 game in 2012 using HDInsight to store and process data of all actions by users. They used share point power view to show the digital dashboard for player statistics from the HDInsight cluster. They produced reports about the game (playlist used by players, game mode, maps played, weapons used) daily, weekly and monthly. All this information was stored in HDInsight. They used Avro to serialize the raw data into data objects. In the front end, they used SharePoint with power view to work with data and also used Microsoft Excel. Using such data, they discovered the trends to keep the game interesting and made the game evolve better over time. They found what playlist was working, what maps were interesting, and what is not working. Every week they released a new and better playlist and kept it fresh to improve customer satisfaction.

6.24.2   HADOOP ON AWS

6.24.2.1   MANUALLY DEPLOYING HADOOP CLUSTER ON AWS BY CREATING VMS

Major steps involved in creating an EC2 instance in AWS:

1. Create an AWS account
2. Create and launch an EC2 instance
3. Connect the launched VM via putty

1. **Create an AWS account**: need a mail id and credit/debit card
   Visit https://aws.amazon.com/console/ → create a free account

Select "professional" account type and fill up the details

Agree on terms and conditions → Create an account and continue

Keep your credit/debit card ready to provide payment details. AWS will charge 2 rupees (will be refunded back in 10 days after removing your AWS account).

Payment Information

Please type your payment information so we can verify your identity. We will not charge you unless your usage exceeds the AWS Free Tier Limits. Review frequently asked questions for more information.

> (i) As part of our card verification process we will charge INR 2 on your card when you click the "Secure Submit" button below. This will be refunded once your card has been validated. Your bank may take 3-5 business days to show the refund. Mastercard/Visa customers may be redirected to your bank website to authorize the charge.

Credit/Debit card number

| |

Expiration date

| 08 ∨ | 2021 ∨ |

Cardholder's name

| RATHINARAJA J |

Billing address
⦿ Use my contact address

Select secure submit → you will receive OTP in the registered mobile number → you will receive a call to verify user registration. Now, select your plan (select free for initial testing purpose)

Your account with the free plan is now successfully activated.

Now, click "sign in to the console" and enter the username and password to start launching the services you wanted.

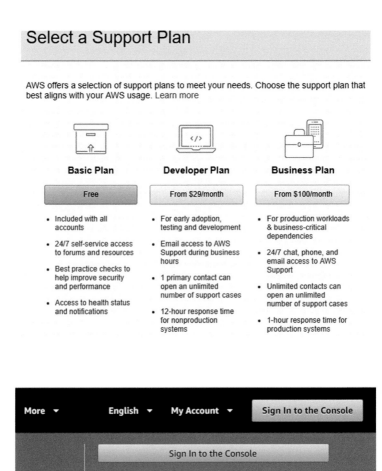

2. Create and launch an EC2 instance

Click on Services

Select EC2 to create a VM (instance)

Select launch instance → you will see a list of OS flavors with different features

Select Ubuntu server 64 bit (free tier instance).

Select only free tier OS flavor (else you will have to pay). If you want scalable instance, select T type instance. Other instance types are fixed resource capability.

Filter by:	All instance types ˅	Current generation ˅	Show/Hide Columns			
Currently selected: t2.micro (Variable ECUs, 1 vCPUs, 2.5 GHz, Intel Xeon Family, 1 GiB memory, EBS only)					
	Family ˅	Type ˅	vCPUs ⓘ ˅	Memory (GiB) ˅	Instance Storage (GB) ⓘ˅
☐	General purpose	t2.nano	1	0.5	EBS only
■	General purpose	t2.micro Free tier eligible	1	1	EBS only
☐	General purpose	t2.small	1	2	EBS only

Click next to configure instance details such as the number of instances to set up multi-node. Let us give 1 for our experiment.

Number of instances ⓘ	1	Launch into Auto Scaling G
	You may want to consider launching these instances into future. Learn how Auto Scaling can help your application	
Purchasing option ⓘ	☐ Request Spot instances	
Network ⓘ	vpc-ebc5f183 (default) ˅	C
Subnet ⓘ	No preference (default subnet in any Availability Zon ˅	
Auto-assign Public IP ⓘ	Use subnet setting (Enable) ˅	

Click next to add storage (for free version only 8 GB is provided). If you want more storage for VM, you will have to pay for it.

Step 4: Add Storage

Your instance will be launched with the following storage device settings. You can attach additional EBS volumes and instanc edit the settings of the root volume. You can also attach additional EBS volumes after launching an instance, but not instance storage options in Amazon EC2.

Volume Type ⓘ	Device ⓘ	Snapshot ⓘ	Size (GiB) ⓘ	Volume Type ⓘ
Root	/dev/sda1	snap-015e0c9bfb72bf22e	8	General Purpose SSD (GP2)

Add New Volume

Click next to add tag (name of instances)

Key	(127 characters maximum)		Value	(255 ch
raja				

Add another tag (Up to 50 tags maximum)

Click next to configure security group → here you can open a list of ports to access on the Internet.

Assign a security group: ⦿Create a **new** security group
◯Select an **existing** security group

Security group name: launch-wizard-1

Description: launch-wizard-1 created 2018-07-27T17:17:32.019+05:30

Type ⓘ	Protocol ⓘ	Port Range ⓘ	Source ⓘ
SSH ⌄	TCP	22	Custom ⌄ 0

Add Rule

Click next to review instance details

▼ AMI Details

⊚ Ubuntu Server 16.04 LTS (HVM), SSD Volume Type - ami-5e8bb23b

Free tier eligible Ubuntu Server 16.04 LTS (HVM),EBS General Purpose (SSD) Volume Type. Support available from Canoni
Root Device Type: ebs   Virtualization type: hvm

▼ Instance Type

Instance Type	ECUs	vCPUs	Memory (GiB)	Instance Storage (GB)
t2.micro	Variable	1	1	EBS only

▼ Security Groups

Select launch

Once launched, you will be prompted to create a private key to log in remotely with your VM.

Select create a new key pair and enter a key name (your choice) → select to download that key. Key is in .pem format. Select launch instances.

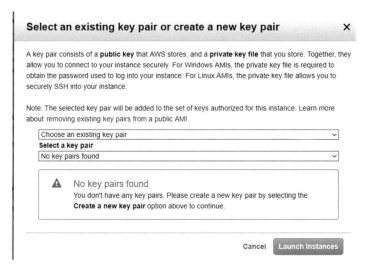

You can see the status of VM launched.

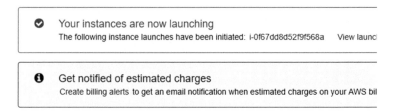

Go to → services → compute → EC2 to see running/paused instances

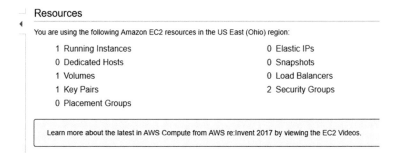

Select running instances → you can see its properties, IP address, port numbers opened

3. Connect the launched VM from your machine

From Windows machine: by default, Ubuntu VM contains an SSH server for remote communication. Windows do not contain SSH client to connect with an SSH server. Therefore, we use a software called "putty" in Windows, which contains an SSH client.

Launch Instance ▼	Connect	Actions ▼					
Q Filter by tags and attributes or search by keyword							
☐ Name ▼	Instance ID ▲	Instance Type ▼	Availability Zone ▼	Instance State ▼	Status Checks ▼	Alarm Statu
☐	i-0f67dd8d52f9f568a	t2.micro	us-east-2c	● running	✅ 2/2 checks ...	None

Instance: | i-0f67dd8d52f9f568a   Public DNS: ec2-18-217-173-239.us-east-2.compute.amazonaws.com

Description	Status Checks	Monitoring	Tags

Instance ID	i-0f67dd8d52f9f568a	Public DNS (IPv4)	ec2-18-217-173-23 2.compute.amazon
Instance state	running	IPv4 Public IP	18.217.173.239
Instance type	t2.micro	IPv6 IPs	-
Elastic IPs		Private DNS	ip-172-31-38-216.u:
Availability zone	us-east-2c	Private IPs	172.31.38.216
Security groups	launch-wizard-1 . view inbound rules . view outbound rules	Secondary private IPs	
Scheduled events	No scheduled events	VPC ID	vpc-ebc5f183

In order to remotely connect the VM via putty in Windows, Use puttygen software to convert .pem format key to .ppk format. Because putty understands only .ppk for remote connection. So, download puttygen from https://www.chiark.greenend.org.uk/~sgtatham/putty/latest.html

pageant.exe (an SSH authentication agent for PuTTY, PSCP, PSFTP, and Pl

32-bit: pageant.exe (or by FTP)
64-bit: pageant.exe (or by FTP)

puttygen.exe (a RSA and DSA key generation utility)

32-bit: puttygen.exe (or by FTP)
64-bit: puttygen.exe (or by FTP)

putty.zip (a .ZIP archive of all the above)

32-bit: putty.zip (or by FTP)
64-bit: putty.zip (or by FTP)

Run puttygen.exe

Click "load" to upload the downloaded .pem file → select save the private key → give the same name of the .pem file → close puttygen.

Download putty from https://www.chiark.greenend.org.uk/~sgtatham/putty/latest.html and launch.

Select auth in the category section → browse for a .ppk file to select

Select Session → enter public IP of launched VM.

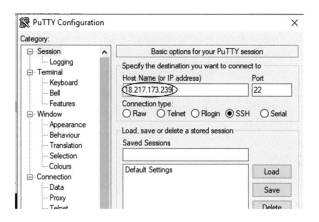

You can find the public IP of launched VM from running instances dashboard.

Enter user name "ubuntu" and there is no password by default.

Now, you are accessing VM running in a remote cloud data-center as you have it locally in your computer. Similarly, you may create multiple VMs and set up a cluster for Hadoop. However, to set up a multi-node cluster, first, create a network and then create multiple VMs. While creating VMs, add VM to that particular network. This will form a private virtual cluster of VMs. While configuring VM, you have to open the particular Hadoop port numbers in the security group configuration.

6.24.2.2   CREATE AND LAUNCH EMR

It is a manual work creating a set of VMs, adding to a private network, and setting up Hadoop. However, AWS provides Hadoop itself as a service via Elastic MR. Therefore, in a few minutes, you can set up any size of Hadoop cluster you want. AWS offers Hadoop as a service on two different platforms: a cluster of VMs or a cluster of PMs. We are going to get EMR with a cluster of VMs. Please note that EMR is not a free service, you will be charged as much as you consume.

Services → Analytics → select EMR

Select create cluster → you can give a cluster name and select S3 storage, MR or Spark, specify how many VMs you want.

General Configuration

Cluster name | My cluster

☑ Logging ❶

S3 folder | s3://aws-logs-585869820034-us-east-2/elasticm

Launch mode ● Cluster ❶   ○ Step execution ❶

Software configuration

Release | emr-5.16.0 | ❶

Applications ● Core Hadoop: Hadoop 2.8.4 with Ganglia 3.7.2,
Hive 2.3.3, Hue 4.2.0, Mahout 0.13.0, Pig 0.17.0,
and Tez 0.8.4

○ HBase: HBase 1.4.4 with Ganglia 3.7.2, Hadoop
2.8.4, Hive 2.3.3, Hue 4.2.0, Phoenix 4.14.0, and
ZooKeeper 3.4.12

○ Presto: Presto 0.203 with Hadoop 2.8.4 HDFS and
Hive 2.3.3 Metastore

○ Spark: Spark 2.3.1 on Hadoop 2.8.4 YARN with
Ganglia 3.7.2 and Zeppelin 0.7.3

Create cluster → Hadoop cluster is launched within a few minutes. You cannot edit cluster

Hardware configuration

Instance type | m4.large | The selected instance type adds a default EBS volume per instance. Learn more

Number of instances | 3 | (1 master and 2 core nodes)

Security and access

EC2 key pair | Choose an option | ❶ Learn how to create an EC2 key pair.

Permissions ● Default ○ Custom
Use default IAM roles. If roles are not present, they will be automatically
created for you with managed policies for automatic policy updates.

EMR role EMR_DefaultRole ❶

EC2 instance profile EMR_EC2_DefaultRole ❶

configurations as it is internally provided. Go to → analytics → EMR to see running clusters.

6.24.2.3   *STOP THE RUNNING SERVICES AND REMOVE YOUR ACCOUNT*

When you do not use EC2 instance or EMR service, you have to stop them. Because free usage with EC2 services is limited. If you do not stop the services which you do not use, you will be charged accordingly from your credit/debit card.

Before removing your AWS account, you have to stop all the running services from your AWS account.

Go to → Services → Compute → Running Instances → Actions → Instance State → Stop

To close AWS account, go to https://aws.amazon.com/premiumsupport/knowledge-center/close-aws-account/

Follow the procedure given in the following figure.

Once successfully closed your account, it will take up to 10 working days to get the refund 2 rupees.

KEYWORDS

- Hadoop 2.7.0 single node and multi-node implementation
- Hadoop on cloud (Azure and AWS)
- HDFS federation
- HDFS meta-data
- MapReduce and HDFS properties
- MapReduce with Eclipse
- NameNode high availability

CHAPTER 7

Data Science

Your best teacher is your last mistake.
—Abdul Kalam

INTRODUCTION

E-science is an interdisciplinary scientific paradigm that uses large-scale IT infrastructure to process big data. Every business sector and research field are troubled by data in a different fashion for different purposes. For the data to be useful, knowledge (solution) should be extracted for a well-defined problem. In the big data world, finding an approximate solution to the right problem is better than an exact solution to an approximate problem. Moreover, the right problem requires an adequate amount of data to extract good enough solution that can be used for decision making. Data science is, therefore, becoming an essential element for all modern interdisciplinary research, to facilitate collaborative scientific discovery and involve the whole life cycle of data analysis.

7.1 BIG DATA RESEARCH AREAS AND JOB ROLES

In general, there are four major research areas in big data: big data analysis, big data infrastructure, big data security, and big applications (as shown in Figure 7.1). Let us discuss these research areas along with the possible jobs roles an employee can acquire there. There are different job roles related to Hadoop like Hadoop developer, Hadoop architect, Hadoop engineer, Hadoop administrator, Hadoop application developer, data analyst, data scientist, business intelligence architect, big data engineer, infrastructure architect, ETL developer/architect, DevOps engineers, etc. We will discuss some typical job roles and their requirements in this section.

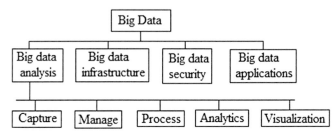

FIGURE 7.1   Big data research area.

7.1.1   *BIG DATA ANALYSIS*

As we discussed already, big data analysis refers to a sequence of steps (capture, store, manage, process, perform analytics, visualize/interpret, and understand) carried out to discover unknown hidden pattern/trend/relationship/association and other useful, understandable, and actionable information (insight/knowledge) that leads to decision making. There are a variety of tools for each step in big data analysis. Refer Section 1.7 to understand the difference between data analysis and data analytics.

1. Data capturing – devices such as sensors, bar code readers, digital camera, etc. capture data in different (structured/unstructured) format.
2. Data storage – captured data may be stored in structured/unstructured databases, files, etc. in HDD, SSD.
3. Managing data – HDFS is used to read/write data from/to disks. Such stored data may be rearranged/deleted depending on the storage capacity available for future data.
4. Processing data – tools like MapReduce, Spark, Pig, Hive, etc. are used to code logics/concepts to process big data in a different fashion (batch/stream/iterative/interactive processing).
5. Performing analytics – applying logics to extract information from big data using tools like statistics, probability, mathematics (calculus, linear algebra, etc.), data mining, machine learning, neural network, optimization algorithm, etc. is called analytics.
6. Visualization/interpretation – extracted information must be presented as graphs/charts to the audience using big data visualization tools like Excel, D3, Tableau, etc. Microsoft Excel is capable of plotting data that is readily available in memory.
7. Understanding data – from the information presented, the audience/analyst arrives in a conclusion, that is decision making.

Data science

Science is all about observing an entity, perform analytics, and understand something from it. This is applicable regardless of the discipline. Similarly, data science deals with the complete cycle of big data analysis. Nonetheless, data science majorly focuses on big data analytics in a comprehensive way to extract useful information. However, data science gives a gentle touch upon all the steps in big data analysis but gives more emphasis on big data analytics. Typical data scientists construct algorithms that are effective and efficient to solve business/research problems and provide new opportunities for decision making from big data. Data science roughly involves several disciplines such as statistics, probability, mathematics (calculus, linear algebra, etc.), data mining and machine learning, distributed computing, optimization algorithms, etc. to transform logic into algorithms (models) to extract useful information from big data.

Data scientist

A person possessing a skill-set of one or more above mentioned disciplines for building efficient, distributed, and scalable algorithms is called a data scientist. A data scientist is typically a domain expert and must have the capability to understand the nature of data and find suitable algorithms for specific problems. Data scientists must have both engineering and analytical skills. They should also be well-versed in creating charts and graphs using visualization tools. Data scientists typically build data analytics libraries such as Mahout, MLBase, etc. to ease others to work with algorithms. In general, Ph.D. is a requirement for typical data scientists along with the skill-set mentioned above. A good data scientist will not just address business problems. Instead, he/she will pick the right problems that have more value to the organization.

Data analyst

Data scientists build data analytical tools (libraries) such as Mahout, MLBase, etc. Data analysts make use of these tools to process data and prepare decent reports. Frameworks like R, Weka, SAS, SPSS, Matlab, etc. include many statistical, data mining, and machine learning libraries built-in. Nowadays, python libraries simplify all these algorithms and are widely used by researchers across disciplines as it is easy to use. The primary task of a data analyst is to prepare required data from DWH (or any other massive corpus of data), feed them to analytical tools, and display the output as easy-to-understand graphs/charts. They usually have computer science and/or business degrees. This profile has nothing to do with big data in particular. A decent mid-sized organization can have many data analysts. Example: sales analyst, marketing analyst. A sales analyst looks at all the sales in the past quarter and figures out a proper sales strategy (whom to target to maximize profits). They will then communicate the report to the leader/boss. Data analyst looks at only data from a single source, whereas data scientists will more likely explore and examine data from multiple disparate sources for building algorithms (models).

7.1.2 BIG DATA INFRASTRUCTURE

Building big data infrastructure (frameworks and tools, please refer Section 1.9) to support data analysis is another important research area. Without big data infrastructure, it is impossible to implement and test the logic on big data. A key property to hold while building big data infrastructure is to improve scalability, fault tolerance, and energy efficiency. Cloud computing based big data infrastructure is increasing nowadays to acquire the advantages of cloud computing features.

Big data engineer

A data scientist creates insights, while a data engineer creates things. This job role is all about working in the data-center and handling storage devices for gathering managing data. They need to understand the hardware, especially storage architectures (SAN, NAS, DAS). A good data engineer should possess extensive knowledge of database and best engineering practices. These include handling logging errors, monitoring system, scaling storage and integration, knowledge of database, back up, etc. Data engineers need not hold a Ph.D. but should have knowledge and expertise in one or more database software (SQL/NoSQL/DWH). Data scientists also should have some understanding and working skill-set of data engineering, whereas a data engineer need not have the skill set of data scientists.

Big data architect

Large enterprises generate a huge amount of data from different sources such as CRM, stock market, etc. A data architect is someone who can understand all the sources of data and work out a plan for integrating and maintaining all the data centrally. This includes designing a database, developing strategies for data acquisitions, achieve recovery, implementation of a database, cleaning and maintaining databases, etc. In computer science, data architecture is composed of models, policies, rules, or standards that govern which data should be collected, how it is stored, arranged, integrated, and put to use.

7.1.3 BIG DATA SECURITY

Medical records, customer transaction details, etc. are highly sensitive and should be protected from breach. Providing security to large scale data is not a simple task. Encrypting huge data is compute-intensive and time-consuming process while reading and writing. Moreover, promising privacy is also to be considered. Nowadays, blockchain is emerging and becoming an interesting concept for achieving data security.

7.1.4 BIG DATA APPLICATIONS

Big data is prevalent and exists in various disciplines (Engineering, Medicine, Law, Science, Finance, Education, Healthcare, etc.). For instance, in a smart city, there are different scenarios to handle big data and come up with a better decision. Computer science and information technology background researchers are good at using big data frameworks/tools for building algorithms, but they lack domain knowledge. Consider a researcher in the medical field who is a domain expert but do not have skill-set to work with tools. So, establishing inter-disciplinary research will improve life easier.

7.2 BIG DATA ANALYTICS

As we discussed enough already, applying logic to extract information from big data is called big data analytics, which is what data science covers elaborately. Data analytics differ concerning different types of data and the level of intelligence required. An algorithm designed for social media applications (audio/video dataset) may not be suitable for healthcare datasets (highly inter-linked text dataset). Similarly, the level of intelligence required to extract from dataset also differs across applications. For instance, the intelligence required for playing chess game is typically high than other scientific applications like weather simulation, etc. Considering this, let us discuss different big data applications and its information extraction by applying data analytics. For instance, if data analytics is applied in text data, it is called text analytics, and so on.

7.2.1   TEXT ANALYTICS

Text analytics helps to extract information from unstructured text data. It is applied on social network data such as tweets, Facebook comments, and chats, emails, blogs, online forums, survey responses, corporate documents, news, call-center logs, etc. Understanding the semantics of text is called computational linguistics, also known as Natural Language Processing (NLP). This is also called text mining. Text analytics generate meaningful summaries and insights from large text dataset to help businesses for evidence-based decision-making. Example: movie success/failure prediction from comments, stock market prediction from financial news. There are different applications of text analytics: information extraction, text summarization, question answering, opinion mining.

7.2.1.1   INFORMATION EXTRACTION

Deriving structured information from unstructured data is called information extraction. Example: drug name, drug dosage, and its frequency discovery from medical prescriptions. Primary sub-tasks of information extraction are entity recognition and relation extraction.

- Entity recognition searches nouns, numeric in texts, and segregates into predefined entities such as name, date, location, organization, etc.
- Relation extraction explores and constructs semantic relationships among entities found in the entity recognition.
- Example: C was introduced in 1979 by Dennis Ritchie, USA.
- Entity recognition → 1981 as the year, C and Dennis Ritchie as name, the USA as place.
- Relation extraction → place of birth of C is the USA.

7.2.1.2   TEXT SUMMARIZATION

Generating a summary of one or multiple documents is called text summarization. The generated summary must be able to deliver useful and important information from original texts and documents. Applications include scientific journals, news articles, advertisements, emails, and blogs. Text summarization employs two methods: extractive and abstractive methods. In extractive summarization, a summary (usually sentences) is generated from the given dataset. Summary of the result is a subset of the original document. Extractive summarization need not deliver any meaning and understanding of the original text and documents. Abstractive summarization helps to extract semantic information from the text. Texts present in the resulting summary need not necessarily be a subset of the original documents. Advanced NLP methods are mostly used to extract abstractive summary from the original text. However, extractive systems are more comfortable to adopt, especially for big data.

7.2.1.3 QUESTION ANSWERING (QA)

Questions raised in natural language can be answered using these techniques. Many commercial QA systems are faster and answer meaningfully. Example: Apple Siri and IBM Watson. Healthcare, marketing, finance, education, etc. are some of the application areas of QA systems. QA systems highly rely on NLP techniques similar to abstractive summarization. There are three categories in QA techniques:

- Information retrieval-based approach
- Knowledge-based approach
- Hybrid approach.

Information retrieval – based QA systems has three sub-components.

Question processing – determines the type and focus of the question, and predicts the answer type. It also finds the tenses of question and answer.

Document processing – Based on the question type and focus from question processing, relevant pre-written texts are retrieved from a set of existing documents.

Answer processing – There can be many relevant passages for a question type. Some of the more suitable answers, called candidate answers, from those relevant passages, are selected.

Knowledge-based QA systems generate a syntactical form from the question, using which structured resources are queried. Typical application areas are transportation, tourism, medicine, etc. where very less pre-written documents are available. Apple Siri QA system exploits the knowledge-based approach. In hybrid QA systems, rather than analyzing questions semantically, candidate answers are extracted using information retrieval methods. IBM Watson uses this approach.

7.2.1.4   SENTIMENT ANALYSIS (OPINION MINING)

Everybody has started to use smartphones. People share their opinions towards entities such as products, organizations, events, individuals, etc. in social media. It is a good source of data to find what group of people think about a particular entity. Some of the application areas are marketing, finance, political, and social events. There are three ways to perform sentiment analysis: document-level, sentence-level, aspect-based.

Document-level techniques identify whether the whole document conveys negative or positive opinion about a single entity. Some of the techniques classify a document into positive or negative while most of the techniques include more sentiment classes (like Amazon's five-star system). Sentence-level techniques find the polarity of a single sentiment about a known entity. Subjective sentences are distinguished from objective ones using sentence-level techniques. Aspect (feature) based techniques identify different sentiments possible in a document and find the aspect of the entity to which each sentiment refers. Example: product and film reviews.

7.2.2   AUDIO ANALYTICS

Extracting information from unstructured audio data is called audio analytics. Applying this to human speech recognition is called speech analytics. Some of the application areas are call-center, healthcare, etc. where millions of recorded calls are analyzed for threat detection, improving customer satisfaction, finding customer behavior, identifying service issues, etc. Live calls also can be analyzed and given feedback in real-time. Also, automated call-center use interactive voice response platforms to identify and handle frustrated callers.

In healthcare, diagnosing medical conditions from patients talking pattern, analyzing infants cry (contain information about the infant health and emotional status) are the better use cases of audio analytics. There are two conventional approaches followed in speech analytics: Transcript-based and phonetic-based. The first approach consists of two-phases: indexing and searching. In the indexing phase, the speech content of the audio is transcribed. It is performed using automatic speech recognition algorithms that match the sounds to words, which are then identified based on a predefined dictionary. It generates a similar word if the system fails to find the exact word in the dictionary. The output of the system is a searchable index file that contains information about the sequence of the words spoken in the speech. In the searching phase, standard text-based methods are used to find the search term in the index file.

Phonetic-based systems deal with sounds or phonemes. Phonemes are the perceptually distinct unit of sounds in a specified language that distinguishes one word from another. Phonetic-based systems also consist of two phases: phonetic indexing and searching. In the first phase, the system translates the input speech into a sequence of phonemes. This is in contrast with large vocabulary continuous speech recognition systems where the speech is converted into a sequence of words. In the second phase, the system searches the output of the first phase for the phonetic representation of the search terms.

7.2.3 VIDEO ANALYTICS

The process of extracting information from video data is called video analytics or video content analysis. It is done on pre-recorded videos and real-time video. Increasing usage of CCTV produces abundant video data which requires information on real-time. Pre-recorded videos in video sharing community (YouTube, Netflix) are analyzed to improve customer experience by suggesting interesting videos. Massive amount of video data takes more time process and comes up with a decision. Example: one second of a high-definition video, in terms of size, is equivalent to over 2000 pages of text. How about processing huge videos and categorizing them into different titles on YouTube?

Automated security and surveillance camera in restricted areas such as military, forest, etc. are some of the primary application areas of video analytics in recent years. Upon detection of a threat or breach, the surveillance system may notify security personnel in real time or trigger an automatic action (Example: sound alarm, lock doors, or turn on lights). It can largely minimize the labour cost. CCTV data in retail outlets help to place products in the right place where customers mostly spend their time in the supermarket. Retailers can count the number of customers, measure the time they stay in the store, detect their movement patterns, measure their dwell time in different areas, monitoring queues in real time, etc.

Video indexing can be done using meta-data of the video, soundtrack, or transcripts. RDBMS is used to store meta-data and search the videos based on data. Video analytics is done in two different places: centralized processing, or edge processing. Captured video data from CCTV/sensors is stored in a centralized data-center. Data from edge devices is compressed before sending over the Internet as public bandwidth is costly. Therefore, we lose some data before we process; this affects the accuracy in the outcome. In edge processing, analytics is applied in the device, which generates video data. So, the entire video data is processed without losing tiny content. However, processing in the edge devices itself is costlier than central processing approach as it needs more power and takes more time.

7.2.4   SOCIAL MEDIA ANALYTICS

Social media such as Facebook, Twitter, blogs, etc. is an effective way of communication medium at present, and people contribute huge data nowadays by sharing their feedback, views/thoughts, reviews, emotions, etc. as for comments. Different opinions on different entities help to identify what people estimate about that particular entity. The research on social media analytics broadly covers several disciplines, including bio-informatics, psychology, sociology, and anthropology. Marketing has been the primary application of social media analytics in recent years. User-generated content (texts, images, videos, and bookmarks), relationships and interactions between the network entities (people, organizations, and products) are the primary sources of information in social media. Social media analytics can be classified into two groups: content-based analytics and social network analysis.

7.2.4.1   CONTENT-BASED ANALYTICS

It focuses on the data posted by users on social media platforms. It is possible to perform text analytics, audio analytics, video analytics, etc. from such social media data, which is often voluminous, unstructured, noisy, and dynamic.

7.2.4.2   SOCIAL NETWORK ANALYSIS (SNA)

This is also called as structure-based analytics and deals with mapping relationships and flows between the entities (denoted as nodes in the graph).

Nodes in the network denote people and groups while the links denote the relationships or flow between the nodes. SNA helps to predict the links (line prediction) among nodes, and predict communities or hubs (nodes having a large number of connections). Example: How many degrees of distance are you from MS Dhoni on Facebook? SNA has also been applied in various applications such as anthropology, bio-informatics (protein structure evaluation), communication studies, economics, geography, history, information science, organizational studies, social psychology, etc.

Analysis of networks consisting of millions of connected nodes is computationally costly. However, social computing is a hot cake at present in the Internet-based companies. Processing high dimensional data was already a tough task in current scientific research. Early, we did dimensionality reduction (using PCA, LLE, etc.) with less loss of information as possible. Remember, big data should not be restricted in minimizing dimension.

7.2.5 *PREDICTIVE AND PERSPECTIVE ANALYTICS*

As shown in Figure 7.2, upon having past data, if we can analyze and report what and why something has happened, it is called pattern mining. If analytics is done on real-time streaming data and gives a result within a deadline, it is called real-time analytics. Upon having past and present data, if we can predict what will likely to happen in the future, it is called predictive analytics. If analytics predict something and instruct us what to do in the future, it is called perspective analytics. In general, these can be applied to almost all disciplines. Example: jet engines have thousands of sensors producing a huge stream of data. Using such data, we can predict the failure of jet engines. Similarly, based on the items kept in a basket, we can predict what customer may like to buy now with the help of real-time data.

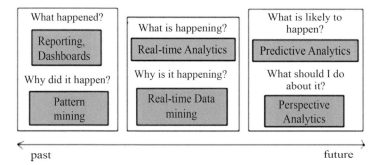

FIGURE 7.2 Big data research area.

7.2.6   *GRAPH PROCESSING*

Graphs are just a type of data modeling are a representation that provides a flexible abstraction for describing the relationships among a discrete set of objects using nodes and edges. Many real-world problems can be modeled as graphs and solved using graph algorithms. The large, complex graph may look as shown in Figure 7.3. For example, Facebook users are nodes, and friendship/relationship is the edge.

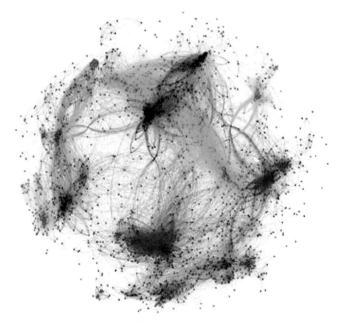

FIGURE 7.3   Big graph.

With no doubt, a large graph needs a large-scale platform because large graph does not fit into the memory of a single computer. Hadoop is suitable for large scale graph processing, but it is slow due to huge dependencies and multiple iterations. Hadoop mostly concentrates on record-level parallelism. However, graph data is based on nodes and edges (not like records), which show requires iterative computation and communication due to large-scale dependencies. Therefore, a lot of graph-based frameworks/tools emerge to optimize the performance of large-scale graph data processing. Example: Pregel, Giraph, GraphX, GraphLab, PowerGraph, Graphchi, etc.

Large scale graph processing is highly challenging due to the following reasons.

1. Unstructuredness:
   - It is challenging to extract parallelism based on the partitioning of the data due to the irregular structure of graphs.
   - It limits the scalability of processing because of unbalanced computational loads resulting from poorly partitioned data.
2. Data-driven computations
   - Nodes and links of the graph dictate the computations.
   - It is difficult to express parallelism based on the partitioning of computation because the structure of computations (algorithm steps) is not known a priori.
3. Poor data locality: The computations and data access patterns do not have many localities due to the irregular structure of graphs.
4. High data access to computation ratio
   - Graph algorithms are often based on exploring the structure of a graph.
   - The runtime can be dominated by waiting memory fetches due to poor data locality.

KEYWORDS

- **big data job positions**
- **data science**

APPENDIX A

Public Datasets

Most of the time, it is not possible to get already accumulated big data for specific applications, so we use benchmark programs [21] to generate huge data for experiments. However, from websites [7], [8], you can download some publicly available big dataset. Here, we list some public repositories to download big data.

Datasets of various applications

https://www.kdnuggets.com/datasets/index.html
https://www.kaggle.com/datasets
https://knoema.com/
http://aws.amazon.com/datasets
https://datasource.kapsarc.org/pages/home/
https://www.kaggle.com/datasets
https://data.austintexas.gov/
https://data.cityofchicago.org/
https://www.govloop.com/
https://data.gov.uk/
https://data.medicare.gov/
https://data.seattle.gov/
https://datasf.org/opendata/
https://ftp.ncbi.nih.gov/
https://www.tylertech.com/products/socrata
https://www.archives.gov/research
https://pslcdatashop.web.cmu.edu/
https://www.data.gov/
https://www.imdb.com/interfaces/
https://datacite.org/

Books

http://gutenberg.org
http://cbt.gg/18yYJHn

Wikipedia

http://en.wikipedia.org/wiki/wikipedia:database_download

Machine learning

http://archive.ics.uci.edu/ml/index.php
http://sci2s.ugr.es/keel/index.php
http://alias-i.com/lingpipe/demos/tutorial/classify/read-me.html
http://www.datasciencetoolkit.org/developerdocs

NLP

http://www.nltk.org/
https://github.com/nltk/nltk

Graph

http://web.stanford.edu/class/cs224w/data.html

Government

http://hdr.undp.org/en/statistics/data/

Weather

http://www1.ncdc.noaa.gov/pub/data/normals/1981-2010/source-data-
sets/isdlite-normals.tar.gz
ftp://ftp.ncdc.noaa.gov/pub/data/uscrn/products/daily01/

Finance

http://eoddata.com/
https://www.assetmacro.com/
https://www.globaldata.com/timetric/
https://www.bls.gov/data/
https://opendata.cityofnewyork.us/

NASA

https://www2.jpl.nasa.gov/srtm/cbanddataproducts.html

Cross-disciplinary data repositories like medical, space, etc.

https://www.mrlc.gov/data?f%5B0%5D=category%3Acanopy
https://www.faa.gov/data_research/
https://www.dartmouthatlas.org/
http://usgovxml.com
https://registry.opendata.aws/
https://datacite.org/
https://figshare.com/
http://linkeddata.org
https://data.opendatasoft.com/pages/home/
http://thewebminer.com/
https://www.quandl.com/

Social network datasets

http://enigma.io

Webservice-related repository

http://build.kiva.org/

Network data repository

http://networkrepository.com/
http://crawdad.org/
https://www.factual.com/

Scientific data repository

http://mlvis.com/

APPENDIX B

MapReduce Exercise

We have given a bunch of exercises to help you work with map/reduce tasks and understand the input and output of map/reduce tasks. Java codes and discussions for all these exercises are given in [17]. You can download and work in parallel while reading the book chapters.

Exercise 1: Practice with Linux commands, Java basic, and collections.

Exercise 2: Set up single/multi-node Hadoop or use it in Eclipse itself (refer Chapter 3, and Section 6.1 to Section 6.4).

Exercise 3: Work with HDFS and YARN commands in detail (refer Section 6.6).

Exercise 4: Commissioning and decommissioning nodes (refer to Section 6.3).

Exercise 5: Write a wordcount program in C and Java to find the number of occurrences of each word in a huge file (100 MB or beyond). Observe the running time in both languages.

    `WCJava.java` in `practice` package

Exercise 6: Write a wordcount MR job using HDT.

    Writing MR wordcount job using Eclipse (refer Section 6.5).

Write a wordcount job in MRv1 (all in one, and separately) using MRv1 package (refer Section 3.4).

    `WCMRv1AllinOne.java` in `MRV1` package (all in one)

    `WCMRv1Driver.java` in `MRV1` package

    `WCMRv1Mapper.java` in `MRV1` package

    `WCMRv1Reducer.java` in `MRV1` package

Write a wordcount job in MRv2 (all in one, and separately) using MRv2 package (refer Section 6.5.1).

    `WCMRv2AllinOne.java` in `MRV2` package (all in one)

    `WCMRv2Driver.java` in `MRV2` package

    `WCMRv2Mapper.java` in `MRV2` package

    `WCMRv2Reducer.java` in `MRV2` package

Launching MR wordcount job from Eclipse (on Ubuntu) onto Hadoop using HDT (refer Section 6.5.2).

Run a simple wordcount job in Eclipse itself on Ubuntu/Windows (refer Section 6.5.3).

Exercise 7: Launch an MR wordcount job on the command line.

Write all the jobs in MRv2 hereafter.

- Using commands to compile, convert into the jar, and run in Hadoop environment.
- Include all the commands in a shell script and launch in Hadoop environment.

Exercise 8: Hadoop Streaming and Pipes (refer Section 6.8.4)

Exercise 9: Launch an MR wordcount job using maven in Eclipse.

Exercise 10: Experience with MapReduce wordcount job for small vs large files.

- Upload a small and large file onto HDFS.
- Show how many blocks are created for each file in WUI with single and multi-node.
- Observe the differences such as running time, the number of map/ reduce tasks, records processed, etc. displayed in the console.

Exercise 11: Write MR jobs for the following. (use the card dataset)

- Display wordcount
- Display word that occurs over 5 times
- Display bigrams
- Display bigrams that occur over 5 times
- Display word that occurs between 5 to 10 times
- Display bigrams that occur between 5 to 10 times

Sample programs

        `CardCountBySuite.java` in `cards` package
        `CardRecordCount.java` in `cards` package

Exercise 12: Exercise 12: Explore different dataset locations (refer to APPENDIX A).

Exercise 13: Browse Hadoop API libraries and grepcode.com (refer to Section 6.9.1).

Exercise 14: Write an MR wordcount job with no mapper and no reducer.

    `NoMapRedDef.java` in `practice` package

Exercise 15: Write an MR wordcount job with default Mapper and Reducer.

    `DefMapRed.java` in `practice` package

Exercise 16: Write an MR wordcount job with the mapper and no reducer.

    `MapNoRed.java` in `practice` package

    `WCZeroRedDriver.java` in `practice` package

Exercise 17: Write an MR wordcount job to sort map output

    `MapSortedOutDriver.java` in `practice` package

Exercise 18: Write a combiner for wordcount job and observe the number of input records in counters for reducer before and after using combiner.

    `WCCombinerDriver.java` in `practice` package

Exercise 19: Write an MR wordcount job with multiple reduce tasks and observe the output files and its input load in counters.

    `WCMultiRedDriver.java` in `practice` package

- 0 reducer
- Default/identity reducer
- More than one reducers

When reduce tasks should be launched (refer Section 6.9.4.3)

Exercise 20: Runtime arguments passing via command-line. Example: find words that occur 2 to 4 times.

Tool vs ToolRunner vs GenericOptionsParser (refer Section 6.7.5 and Section 6.7.6)

    `WCMRv2Driver.java` in `RunTimeArguments` package

    `WCMRv2Mapper.java` in `RunTimeArguments` package

    `WCMRv2Reducer.java` in `RunTimeArguments` package

$ yarn jar Job.jar RunTimeArguments.WCMRv2Driver -D UpperLimit=3 -D LowerLimit=2 /input.txt /output

Exercise 21: Write a MR wordcount job and test with different block size, IS size. Observe the number of map tasks and output files.

    `ISsizeDriver.java` in `practice` package

Exercise 22: Controlling IS size by using NLineInputFormat (fixing number of lines per map task)

`NLineInputDriver.java` in `practice` package
`NLineMapper.java` in `practice` package

Exercise 23: Processing a single file regardless of IS size by a mapper

`WholeFileInputFormat.java` in `practice` package
`WholeFileISDriver.java` in `practice` package
`WholeFileMapper.java` in `practice` package
`WholeFileRecordReader.java` in `practice` package

Exercise 24: Writing user-defined counters and display them in the result

`CounterDriver.java` in `practice` package
`CounterMapper.java` in `practice` package

Exercise 25: Logging

Exercise 26: Too many small files (refer Section 6.9.4.4).

Solution 1: You can modify the IS size to be bigger.

Solution 2: Combine all files manually and then upload onto HDFS

Solution 3: Many files to one IS using CombineTextInputFormat

`CombineFilesDriver.java` in `practice` package
`WCMRv2Mapper.java` in `practice` package
`WCMRv2Reducer.java` in `practice` package

Solution 4: Hadoop archive

Solution 5: SequenceFileInputFormat

`SmallFilesToSequenceFileConverter.java` in `practice` package
`WholeFileInputFormat.java` in `practice` package
`WholeFileRecordReader.java` in `practice` package

Exercise 27: Understanding hash partitioner and writing custom partitioner.

`WCPartitioner.java` in `practice` package
`WCPartitionerDriver.java` in `practice` package

Exercise 28: more sample jobs

Wordcount with no case sensitive
Hadoop example jobs

Exercise 29: Inverted index

`InvertedIndexMapper.java` in `InvertedIndex` package
`InvertedIndexReducer.java` in `InvertedIndex` package
`InvertedDriver.java` in `InvertedIndex` package

Exercise 30: Multiple output file from the map or reduce task

`MultipleOutputFileMapper.java` in `practice` package
`MultipleOutputFileReducer.java` in `practice` package
`MultipleOutputFileDriver.java` in `practice` package

Exercise 31: Understanding and implementing custom key and value for counting the number of occurrences of IP address in the server log.

`CustomKey.java` in `CustomKeyValue` package
`CustomDataTypeMapper.java` in `CustomKeyValue` package
`CustomDataTypeReducer.java` in `CustomKeyValue` package
`CustomDataTypeDriver.java` in `CustomKeyValue` package

Now try using CustomKey as value

Exercise 32: Multiple values in mapper output

`MultiWritableValue.java` in `CustomKeyValue` package
`MultiWritableMapper.java` in `CustomKeyValue` package
`MultiWritableReducer.java` in `CustomKeyValue` package
`MultiWritableDriver.java` in `CustomKeyValue` package

Exercise 33: Creating custom InputFormat for finding out the number of people and their sex who died and survived from Titanic dataset.

Step 1: implement a custom key which comprises 2nd and 5th column (composite key)

`TitanicCustomKey.java` in `Titanic` package

Step 2: create custom InputFormat

`TitanicInputFormat.java` in `Titanic` package

Step 3: implement custom Record Reader

`TitanicRR.java` in `Titanic` package

Step 4: mapper using custom key and value

`TitanicMapper.java` in `Titanic` package

Step 5: reducer using custom key and value

`TitanicReducer.java` in `Titanic` package

Step 6: driver method

`CustomInputFormatDriverTitanic.java` in `Titanic` package

Exercise 34: Chaining mappers.

Job1: wordcount program

job2: counting words that start with the same letter from the output of Job1.

`Mapper1.java` in `ChainingMRjobs` package

`Mapper2.java` in `ChainingMRjobs` package

`Mapper3.java` in `ChainingMRjobs` package

`Reducer1.java` in `ChainingMRjobs` package

`Reducer2.java` in `ChainingMRjobs` package

`ChainMapperDriver` in `ChainingMRjobs` package

Exercise 35: Distributed cache for wordcount job to skip stop-words

1. via MR API in v1 and v2

Write a wordcount job to count words removing stop words like ! , . etc

`DCWCwithStopWordsDriver.java` in `DistriCache` package

`DCWCwithStopWordsMapper.java` in `DistriCache` package

`DCWCwithStopWordsReducer.java` in `DistriCache` package

Working with NCDC dataset

`NCDCDriver.java` in `DistriCache` package

`NcdcMaxTempMapper.java` in `DistriCache` package

`NcdcRecordParcer.java` in `DistriCache` package

`NcdcMaxTempCombiner.java` in `DistriCache` package

`NcdcMaxTempReducer.java` in `DistriCache` package

`NcdcStationMetadataParcer.java` in `DistriCache` package

2. via command line arguments.

$ yarn jar job.jar NYSEDistributedCache.AvgStockVolMonthDriver -files
companylist_noheader.csv /input /out101

Exercise 36: JUnit and MRUnit testing

JunitTest for testing square and counting "a." Adding junit is enough

`JunitTest.java` in `MRUnitTest` package

`JunitSquareTest.java` in `MRUnitTest` package

`JunitTestCountA.java` in `MRUnitTest` package

`AllTests.java` in `MRUnitTest` package

MRUnitTest: run with mrunit-1.0.0-hadoop2.jar and MRv2 jars

`MRDriverUnitTest.java` in `MRUnitTest` package

`MRv2AllinOne.java` in `MRUnitTest` package

`WCMRv2Mapper.java` in `MRUnitTest` package

`WCMRv2Reducer.java` in `MRUnitTest` package

Exercise 37: Switching between local, pseudo-distributed and cluster mode
of Hadoop

hadoop-local.xml
hadoop-localhost.xml
hadoop-cluster.xml

Exercise 38: Understanding and changing YARN Scheduler (refer Section 6.11)

Exercise 39: Compression algorithms

Exercise 40: Tuning MR optimization parameters (refer Section 6.7)

Exercise 41: HDFS federation (refer Section 6.19)

Exercise 42: NN HA (refer Section 6.20)

Exercise 43: Hadoop source to binaries (refer Section 6.18)

Case Study: Application Development for NYSE Dataset

Java codes and discussions for all these exercises are given in [17].

Exercise 1: Get average volume traded for each stock per month from NYSE dataset

1. Develop a parser
    `NYSEParser.java` in `NYSE.Parsers` package
2. Identify input and output formats.
    TextInputFormat and TextOutputFormat as we are dealing with text data.
3. Develop user-defined key and value data types.
    `TextPair.java` in `NYSE.CustomKeyValue` package
    `LongPair.java` in `NYSE.CustomKeyValue` package
4. Develop a mapper
    `AvgStockVolMonthMapper.java` in `NYSE.`
    `AvgStockVolPerMonth` package.
5. Develop a combiner
    `AvgStockVolMonthCombiner.java` in `NYSE.`
    `AvgStockVolPerMonth` package.
6. Develop a reduce task
    `AvgStockVolMonthReducer.java` in `NYSE.`
    `AvgStockVolPerMonth` package
7. Driver function
    `AvgStockVolMonthDriver.java` in `NYSE.`
    `AvgStockVolPerMonth` package

Exercise 2: Working with file system (org.apache.hadoop.fs) APIs

`ListFiles.java` in `FileSystemAPI` package. "input" is an HDFS directory here.

$ yarn jar job.jar FileSystemAPI.GetFiles /input

```
hdfs://node4:10001/input/nyse_2010.csv
hdfs://node4:10001/input/nyse_2011.csv
FilePattern.java in FileSystemAPI package.
```

$ yarn jar job.jar FileSystemAPI.FilePattern /input

`CopyMerge.java` in `FileSystemAPI` package (all small files in a directory are concatenated to a single new file). I created few files in a directory in HDFS.

$ yarn jar job.jar FileSystemAPI.CopyMerge /input /file_name
$ hdfs dfs -cat /file_name

```
URLCat.java
FileSystemCat.java
FileSystemDoubleCat.java
FileCopyWithProgress.java
ListStatus.java
FileDisplay.java
FileDisplay1.java
FileDelete.java
```

Exercise 3: Filtering input files and input records
1. Filtering input files in driver class itself

```
AvgStockVolMonthDriver.java in FileSystemAPI package.
```

2. Filtering input records in the map()
Displaying only BAC stock ticker trade per month in map method itself

```
AvgStockMapperWithProperties.java in Filtering package
AvgStockMapperWithSetUp.java in Filtering package
AvgStockVolMonthDriver.java in Filtering package
```

$ yarn jar job.jar Filtering.AvgStockMapperWithProperties /input /output
$ hdfs dfs -cat /output/part-r-00000 // it displays the stock ticker record of only BAC

3. Filtering input records in the map() via command line arguments

```
AvgStockMapperWithProperties.java in Filtering package
```

Then pass filter.by.stockTicker in command line arguments while launching job. Example:

$ yarn jar job.jar Filtering.AvgStockMapperWithProperties -Dfilter. by.stockTicker=AEO /input /output

$ hdfs dfs -cat /output/part-r-00000 // it displays the stock ticker record of only AEO

4. Filtering input records in setup()

```
AvgStockMapperWithSetUp.java in Filtering package
```

Then pass filter.by.stockticker in command line arguments.

$ yarn jar job.jar Filtering.AvgStockMapperWithSetUp -Dfilter. by.stockTicker=BAC /input /output

$ hdfs dfs -cat /output/part-r-00000 // it displays the stock ticker record of only BAC

$ yarn jar job.jar Filtering.AvgStockMapperWithSetUp -Dfilter. by.stockTicker=BAC,AEO /input /output

$ hdfs dfs -cat /output/part-r-00000 // it displays both AEO, and BAC

Exercise 4: User-defined partitioner

```
TextPair.java in NYSE.partitioner package
//result = prime * result + ((first == null) ? 0 :first.
hashCode());
AvgStockVolMonthMapper.java in NYSE.partitioner package
AvgStockVolMonthCombiner.java in NYSE.partitioner package
AvgStockVolMonthReducer.java in NYSE.partitioner package
AvgStockMonthPartitionerDriver.java in NYSE.partitioner
```
package
```
job.setPartitionerClass(HashPartitioner.class);
```

Upload nyse_data/*.csv files

$ hdfs dfs -put nyse_data /

$ yarn jar Job.jar NYSE.Partitioner.AvgStockMonthPartitionerDriver -D filter.by.stockTicker=BAC,AEO /dataset/nyse_data/*.* /output

All AEO goes to one reducer and all BAC goes to another reducer. Therefore, corresponding output files are

$ hdfs dfs -du /output

```
0 /output/_SUCCESS
0 /output/part-r-00000
```

```
1448 /output/part-r-00001
1568 /output/part-r-00002
0 /output/part-r-00003
```

Exercise 5: Developing custom partitioner

`SecondKeyTextPairPartitioner.java` in `NYSE.partitioner` package

`AvgStockMonthCustomPartitionerDriver.java` in `NYSE.partitioner package`

You can do range-based partitioning, rule-based partitioning, etc.

Exercise 6: Partition decision based on runtime arguments

To choose the way you want partition based on either first field in the custom key or second field in the custom key

`AvgStockMonthCustomPartitionerOptionsDriver.java` in `NYSE.partitioner` package

`FirstKeyTextPairPartitioner.java` in `NYSE.partitioner` package

`SecondKeyTextPairPartitioner.java` in `NYSE.partitioner` package

We can choose the right field that should be used to partition at run time

$ yarn jar job.jar NYSE.Partitioner.AvgStockMonthPartitionerDriver -D partition.by=stockticker /dataset/nyse_data/*.* /newout13

$ yarn jar job.jar NYSE.Partitioner.AvgStockMonthPartitionerDriver -D partition.by=trademonth /dataset/nyse_data/*.* /newout13

You can see some skew in the above output. Check output directory with following command.

$ hdfs dfs -cat /output/part-r-00000 | wc -l

$ hdfs dfs -cat /output/part-r-00001 | wc -l

$ hdfs dfs -cat /output/part-r-00002 | wc -l

$ hdfs dfs -cat /output/part-r-00003 | wc -l

Include the fowling in `FirstKeyTextPairPartitioner.java`

```
partitionValue=new Integer(key.getFirst().toString().
replace("-", "")).intValue()% numReducers;
```

Now, launch the job and check out the results whether all reducers processed and produced equal number of records.

Exercise 7: partitioning and sorting based on the first field in a custom key using a comparator

`AvgStockMonthComparatorDriver.java` in `NYSE.Comparator` package

`AvgStockVolMonthMapper.java` in `NYSE.Comparator` package

`AvgStockVolMonthCombiner.java` in `NYSE.Comparator` package

`AvgStockVolMonthReducer.java` in `NYSE.Comparator` package

`TextPair.java` in `NYSE.Comparator` package

Exercise 8: To get top 3 traded stocks per day from NYSE datasets

`TopThreeStocksByVolumePerDayDriver.java in NYSE.Top3Stocks package`

`TopThreeStocksByVolumePerDayMapper.java in NYSE.Top3Stocks package`

`FirstKeyLongPairPartitioner.java in NYSE.Partitioner package`

`LongPairPrimitive.java in NYSE.CustomKeyValue package`

Try1: Range partitioning to achieve global sorting

`FirstKeyLongPairPartitioner.java in NYSE.Partitioner package`

Try2: To group keys based on the desired field in the composite key

`LongPairPrimitiveGroupingComparator.java in NYSEComparator package`

`job.setGroupingComparatorClass(LongPairPrimitiveGroupingComparator.class);`

To sort keys based on the desired field in the composite key

`LongPairPrimitiveSortingComparator.java in NYSEComparator package`

`job.setSortComparatorClass(LongPairPrimitiveSortingComparator.class)`

Try3: Display top 3 records in each day based on volume in the composite key with user-defined reducer

`TopThreeStocksByVolumePerDayReducer.java in NYSE.Top3Stocks package`

Try4: Global sorting using multiple output file

```
TopThreeStocksGlobalSortingMultipleFiles.java in NYSE.
Top3Stocks package
```

Exercise 9: Compression

Way 1: Specify compression configuration in core-site.xml and mapred-site.xml

Way 2: Specify compression parameters in driver function

```
AvgStockVolMonthCompressionDriver.java in NYSE.
Compression package
```

Run the program for dataset without compression

$ yarn jar job.jar NYSE.AvgStockVolMonthDriver /input /output
$ hdfs dfs -ls /output

```
-rw-r--r-- 3 itadmin supergroup 0 2017-05-20 11:46 /
output/_SUCCESS
-rw-r--r--3 itadmin supergroup 1500726 2017-05-20
11:46 /output/part-r-00000
```

Run the program with compression

$ yarn jar job.jar NYSE.AvgStockVolMonthCompressionDriver /input /
output

Verify the output file `part-r-00000` which is appended by .snappy to denote that it has been compressed.

$ hdfs dfs -ls /output

```
-rw-r--r-- 3 itadmin supergroup 0 2017-05-20 11:46 /
output/_SUCCESS
-rw-r--r-- 3 itadmin supergroup 0726 2017-05-20
11:46 /output/part-r-00000.snappy
```

Moreover, for both cases (compression, without compression) notice the size of output and compare the counters in shuffle and input of reducers to see the difference.

Exercise 10: Counters

You can access counters using three ways: command line, counters API, custom counters (user-defined counter, and dynamic counters).

Way 1: via command line

The general command line syntax is

$ hadoop command [genericOptions] [commandOptions]
$ mapred

$ mapred job

$ mapred job -counter jobid groupname countername

For group name, counter name – browse https://hadoop.apache.org/docs/ r2.7.0/api/

$ mapred job -counter job_1494830873386_0040
        org.apache.hadoop.mapreduce.JobCounter DATA_LOCAL_MAPS

When you execute the above command in Hadoop 2.7.0, JHS must be running if the job is already done. If the job is currently being executed, you need not run JHS to get counters.

Way 2: via counters API in a simple java program

If you want to extract counters from completed jobs using API in the program, then

>   CounterDriver.java in NYSE.Counters package

$ yarn jar job.jar NYSE.Counters.CounterDriver job_1494830873386_0043
        org.apache.hadoop.mapreduce.JobCounter DATA_LOCAL_MAPS

Way 3: custom counters

1. Construct a user-defined counter using an enum to find untraded days (where volume is zero). Enum is defined at compile time itself. Best place to create a user-defined counter is reduce task or after job completion statement in the main method.

    `TopThreeStocksByVolumePerDayReducer.java` in NYSE. Counters package

    `NoTradeDaysCounterDriver.java` in NYSE.Counters package

    `TradeDaysEnum.java` in NYSE.Counters package

    Once the job is done, look at the counters. You will see the user-defined counter name. Example:

    `NYSE.Counters.NoTradeDays`

    `NO_TRADE_DAYS=1`

2. Use a dynamic counter to find untraded days (where volume is zero) at runtime.

    `TopThreeStocksByVolumePerDayReducer.java` in NYSE. Counters package

    `context.getCounter("TradeDaysEnum","NO_TRADE").`
    `increment(1);`

Exercise 11: Joining trade dataset with company header list dataset

`CompanyParser.java` in NYSE.Parsers package

`StockCompanyDistCacheJoinMapper.java` in `NYSE.`
`DistributedCache` package

`StockCompanyJoinDistCacheDriver.java` in `NYSE.`
`DistributedCache` package

Pass NYSE dataset onto HDFS

Keep companylist_noheader.csv dataset in the home folder itself

$ yarn jar job.jar NYSEDistributedCache.AvgStockVolMonthDriver -files companylist_noheader.csv /input /out101

$ hdfs dfs -get /output/part-r-00000 output-part-1 // here I am renaming while downloading

You can set path to cache files. So, you will see filecache, usercache, nmprivate folders in HDFS containing the files we passed.

Web References

1. http://www.go-globe.com/blog/things-that-happen-every-60-seconds/ (accessed on 17 January 2020).
2. https://www.dezyre.com/article/nosql-vs-sql-4-reasons-why-nosql-is-better-for-big-data-applications/86 (accessed on 17 January 2020).
3. A Brief Introduction to Big Data Tools, Bingjing Zhang. https://slideplayer.com/slide/4499827/ (accessed on 17 January 2020).
4. https://github.com/Voronenko/MapReduce-assigments/blob/master/assignment2/src/src/main/BigramCount.java (accessed on 17 January 2020).
5. https://bigdata-madesimple.com/hadoop-market-is-expected-to-reach-50-2-billion-globally-by-2020/ (accessed on 17 January 2020).
6. https://www.marketanalysis.com/?p=349 (accessed on 17 January 2020).
7. https://cloud.google.com/bigquery/public-data.
8. https://www.springboard.com/blog/free-public-data-sets-data-science-project/
9. https://hadoop.apache.org/docs/stable/hadoop-mapreduce-client/hadoop-mapreduce-client-core/MapReduceTutorial.html (accessed on 17 January 2020).
10. https://github.com/gotoalberto/mongo-hadoop-spring-app-example/blob/master/src/main/java/com/gotoalberto/mongohadoopspringappexample/util/WinLocalFileSystem.java (accessed on 17 January 2020).
11. https://hadoop.apache.org/docs/current/hadoop-project-dist/hadoop-common/DeprecatedProperties.html (accessed on 17 January 2020).
12. https://hadoop.apache.org/docs/r1.2.1/mapred_tutorial.html (accessed on 17 January 2020).
13. https://hadoop.apache.org/docs/stable/hadoop-mapreduce-client/hadoop-mapreduce-client-core/MapReduceTutorial.html (accessed on 17 January 2020).
14. https://www.michael-noll.com/tutorials/writing-an-hadoop-mapreduce-program-in-python/ (accessed on 17 January 2020).
15. http://business.fibernetics.ca/blog/scalability-not-thinking-about-it-today-means-youll-pay-tomorrow/ (accessed on 17 January 2020).
16. https://dzone.com/articles/understanding-the-cap-theorem (accessed on 17 January 2020).
17. https://github.com/rathinaraja/MapReduce (accessed on 17 January 2020).
18. https://github.com/rathinaraja/MapReduce/tree/master/dataset (accessed on 17 January 2020).
19. http://ubuntuhandbook.org/index.php/2016/01/how-to-install-the-latest-eclipse-in-ubuntu-16-04-15-10/ (accessed on 17 January 2020).
20. https://github.com/winghc/hadoop2x-eclipse-plugin/tree/master/release (accessed on 17 January 2020).
21. https://engineering.purdue.edu/~puma/datasets.htm (accessed on 17 January 2020).

Index